Finding Your
Seat at the Table

MEDICAL LIBRARY ASSOCIATION BOOKS

The Medical Library Association (MLA) publishes state-of-the-art books that enhance health care, support professional development, improve library services, and promote research throughout the world.

MLA books are dynamic resources for librarians in hospitals, medical research practice, corporate libraries, and other settings. These invaluable publications provide medical librarians, health care professionals, and patients with accurate information that can improve outcomes and save lives.

MEDICAL LIBRARY ASSOCIATION BOOKS PANEL

The MLA Books Panel is responsible for (1) monitoring publishing trends within the industry; (2) exploring new concepts in publications by actively soliciting and proposing ideas for new publications; and (3) coordinating publishing efforts to achieve the best utilization of MLA resources. Each MLA book is directly administered from its inception by the MLA Books Panel, composed of MLA members with expertise spanning the breadth of health sciences librarianship.

ABOUT THE MEDICAL LIBRARY ASSOCIATION

The Medical Library Association (MLA) is a global, nonprofit educational organization, with a membership of more than 400 institutions and 3,000 professionals in the health information field. Since 1898, MLA has fostered excellence in the professional practice and leadership of health sciences library and information professionals to enhance health care, education, and research throughout the world. MLA educates health information professionals, supports health information research, promotes access to the world's health sciences information, and works to ensure that the best health information is available to all.

RECENTLY PUBLISHED MLA BOOKS

3D Printing in Medical Libraries: A Crash Course in Supporting Innovation in Healthcare by Jennifer Herron

Diversity and Inclusion in Libraries: A Call to Action and Strategies for Success Edited by Shannon D. Jones and Beverly Murphy

Framing Health Care Instruction: An Information Literacy Handbook for the Health Sciences by Lauren M. Young and Elizabeth G. Hinton

The Clinical Medical Librarian's Handbook Edited by Judy C. Stribling

The Engaged Health Sciences Library Liaison Edited by Lindsay Alcock and Kelly Thormodson

A History of Medical Libraries and Medical Librarianship: From John Shaw Billings to the Digital Era by Michael R. Kronenfeld and Jennie Jacobs Kronenfeld

Planning and Promoting Events in Health Sciences Libraries: Success Stories and Best Practices Edited by Shalu Gillum and Natasha Williams

Great Library Events: From Planning to Promotion to Evaluation by Mary Grace Flaherty

Assessing Academic Library Performance: A Handbook by Holt Zaugg

Finding Your Seat at the Table: Roles for Librarians on Institutional Regulatory Boards and Committees Edited by Susan M. Harnett and Laureen P. Cantwell

Finding Your Seat at the Table

Roles for Librarians on Institutional Regulatory Boards and Committees

Edited by Susan M. Harnett
and Laureen P. Cantwell

ROWMAN & LITTLEFIELD
Lanham • Boulder • New York • London

Published by Rowman & Littlefield
An imprint of The Rowman & Littlefield Publishing Group, Inc.
4501 Forbes Boulevard, Suite 200, Lanham, Maryland 20706
www.rowman.com

6 Tinworth Street, London SE11 5AL, United Kingdom

British Library Cataloguing in Publication Information Available

Library of Congress Cataloging-in-Publication Data Available

ISBN 978-1-5381-4455-8 (cloth)
ISBN 978-1-5381-4456-5 (electronic)

Contents

Preface

Susan M. Harnett and Laureen P. Cantwell

Service on the institutional review board (IRB) and the Institutional Animal Care and Use Committee (IACUC) is an uncommon activity for academic librarians, despite the fact that many librarians participate in research activities requiring interaction with institutional regulatory boards at their institutions and often conduct original or secondary research as scholars themselves. In a 2012 survey of academic health science libraries, only twenty-two of the 155 Association of Academic Health Sciences Libraries (AAHSL) member institutions surveyed indicated that a librarian was a member of the IACUC at their institution. Of those, only sixteen librarians had full voting privileges.[1] Even less may be known about active participation of librarians on their institution's IRB.[2] While librarians frequently wear many hats in academia, there remains the perception that their role is "limited to that of an information provider, a conduit to resources or even as a gatekeeper,"[3] rather than as academics fluent in research practices and protocols. In 2001, the potential role for librarians' participation within IRBs was highlighted when a study participant at Johns Hopkins University (JHU) died from a contraindication that could have been prevented . . . had the literature review gone back further than 1966.[4] After this tragic incident, JHU's IRB added a librarian to their committee, particularly to assist with thorough literature reviews and literature review assessments. Yet, in the nearly twenty years since, librarians throughout academia still seem to most commonly support clinical and bench science programs and researchers through well-known librarianship services such as literature searching, instruction, and consultations—often without any sort of membership on these key institutional committees.

Despite their fluency in research practices and strategies, the perceived complexity of protocol review and approval in the bench sciences may intimidate a librarian seeking full participation on an IRB or IACUC. Additionally, at the institutional level, a librarian's attaining a seat on such committees will likely require approval by the existing committee leadership and member composition, and perhaps even higher institutional permissions as well. Such conversations will require librarians involved to have a base of knowledge—and an argumentative leg to stand on—in order to secure any spot on the committee, voting or otherwise. Library administration must be prepared to support membership on these committees through professional development opportunities and by capitalizing on previously established relationships with researchers and academic faculty. It may be necessary for library directors to justify the time and effort involved in the librarian's participation, as well as the benefit to the institution as a whole. As health science librarians continue to redefine their roles within the institutional research community, participation on these committees demonstrates the value of the librarian as a partner both in research and in achieving institutional goals.

Librarians who may have an interest in serving on, supporting, or understanding the work of these boards will benefit from developing knowledge in the history and evolution of both human subject and animal research, and an understanding of members' roles and responsibilities. Additionally, librarians will find helpful suggestions for integrating library services into the broader research environment at their institution. This can be done through partnerships, collaborations, and the utilization of skills and knowledge regarding bioinformatics, data science, publishing, and grant identification and writing.

This book provides a background on both human subject and animal research for an understanding of their history and evolution; describes the legal and ethical implications of such research; considers the roles that librarians can play on these boards; and suggests ways to expand the librarian's role on these boards and in overall clinical research engagement. Chapter authors have a variety of critically relevant perspectives as librarians engaged in IRB and IACUC work, including direct experience on these committees as community or nonscientific members, as scientific members, as ex officio (or consultant) members, and in research support-only roles. To bridge the bird's-eye-view chapter content with boots-on-the-ground experiences, this book also features narratives of librarians' experiences with these boards. These accounts not only relate chapter content directly to a librarian-focused audience, but they also paint a picture of what these individuals have learned, how they have used these experiences to market and promote library services to research faculty, and what their visions for the future of librarian-IACUC/ IRB engagement entail.

The book is composed of three sections that will lead readers from more basic levels of understanding of the work of these committees into emerging ways librarians can increase their connection with these important bodies.

Each section will describe the committee in detail—its functions and membership, the duties and responsibilities of those members including protocol review, compliance, and training. They include a short history of the particular area of research, including legal and ethical oversight, federal regulations and policies, and the mission of the regulatory committee or board. Finally, there is a discussion of the ways librarians typically support the work of the boards via research support, literature searching, and best practices. While each section offers unique insights and value to the reader, it should be noted that parts I and II have similar and complementary structures such that succeeding chapters build upon knowledge in previous chapters.

Part I focuses on the work of the IRB, its function, and membership. Chapter 1, by Albright and Harnett, provides fundamental information about the duties and responsibilities of IRB members, the required Collaborative Institutional Training Initiative (CITI) certification, and the practice of protocol review. Highby discusses the evolution of human subject research in chapter 2, providing a meaningful foundation for any librarian discussing original, human subjects research with students or other constituents—as well as contextualizing human subjects research for those in science and medical librarianship roles. From the Hippocratic Oath to the Nuremburg Trials, and from FDA laws and protections to the Belmont Report and the "Common Rule"—seminal documents in the history of human subject research—readers will gain the foundational knowledge essential for active and impactful participation on these committees. This section concludes with chapter 3's attention to best practices for systematic literature reviews. While librarians often have at least some familiarity with literature reviews, and a literature review is not mandatory for human subject research, librarians should understand its potential effect on research and outcomes (e.g., the incident at Johns Hopkins) as well as pro and con arguments regarding whether institutions should enact mandatory literature review policies.

Similar to part I, part II describes the IACUC and its functions in care and use of research animals. Chapter 4 identifies and defines individual members' roles, duties, responsibilities, training, and certification. It discusses the practical implications of membership including protocol review, investigation of noncompliance, facility inspections, reporting, and the types of review. Chapter 5 shares the historical context of animal use in scientific and medical research. It covers the development of legislative oversight by the Office of Laboratory Animal Research (OLAW), Public Health Service (PHS), and those of accrediting committees such as the American Association for Accreditation of Laboratory Animal Care (AAALAC). Next, the Animal Welfare Act and the federally mandated literature review for all sponsored research involving animals—serves as the focus of chapter 6. In this chapter, readers will learn about the three Rs (reduce, refine, replace) as well as effective techniques for performing and evaluating a literature search.

Many librarians possess—and others may choose to pursue—additional training and skill sets in bioinformatics, data management, statistics, and research reproducibility. Other training and development options could focus on grants and grants management, funding source research, copyright, patents, and open access policies and practices. Part III highlights these new and expanding roles for librarians on these boards and in institutional research activities. With this in mind, in chapter 7, Kepsel discusses the art of leveraging relationships with board colleagues into partnerships with researchers and the university. Chapter 8, by Champieux, White, and Alpi, illuminates the needs and training of the undergraduate-on-up researcher in the context of how librarians can use their experience with and on institutional regulatory boards to help upskill students and faculty. Thurman and Savoy take on the unique support needs of veterinary students and faculty in chapter 9, as these programs frequently use live animal models for education and training, which may require alternatives to live animal use as well as consent for client-owned animals. Next, in chapter 10, Nance, Gorman, and Brown detail their work, and the ups and downs of, providing informed consent document review services. Funaro and Nyhan, our chapter 11 authors, explore IACUC reporting guidelines and attach them to library-led instruction and how these guidelines, and teaching about them, can be situated in the larger context of reproducibility librarianship.

Finally, chapter 12 features a variety of narratives from librarians who serve on IRBs and IACUCs as well as librarians who interact with these committees in valuable ways. The goal of this chapter is to provide personal insights into *how* librarians became engaged with their IRB and/or IACUC and the skill sets they bring to the table. For librarians and library administrators curious about IRB and IACUC involvement, these narratives serve to acknowledge roles librarians can play, conversations that may be broached, backgrounds that can be leveraged, the rewards of this kind of institutional service, and more. The editors consider ending the book with these narratives as a compelling way to paint librarians into the institutional regulatory boards and committees, to celebrate the talents and training they connect to their work, and how all of this connects to and benefits library and campus stakeholders. For those librarians interested in serving on an IRB or IACUC, the narratives make excellent cases for engagement that readers can use to connect with their stakeholders. Further, the narratives offer a vision of support services and opportunities librarians can offer without on-committee participation.

Overall, the goal of this book is threefold: to create a thorough knowledge base of information for use by librarians interested in serving on institutional regulatory boards; to suggest additional ways that academic librarians can work with stakeholders to participate in and otherwise support research at their institutions; and to provide librarian- and library-centric contextualization of the realities, challenges, and complexities of this work. This book serves as an aid, then, for those librarians establishing their baseline in this topic, those looking to upskill in order to better aid the work of these committees, and those looking to be a voice of strategic, forward-thinking leadership in their IRB and IACUC communities.

Without further ado, we invite you to take a look at the different seats at the table with these organizations and find *your* seat at the table.

NOTES

1. Susan C. Steelman and Sheila M. Thomas, "Academic Health Sciences Librarians' Contributions to Institutional Animal Care and Use Committees," *Journal of the Medical Library Association* 102, no. 3 (July 2014): 217.

2. Laureen P. Cantwell and Doris Van Kampen-Breit, "Librarians and the Institutional Review Board (IRB): Relationships Matter," *Collaborative Librarianship* 7, no. 2 (2015): 67, https://digitalcommons.du.edu/collaborativelibrarianship/vol7/iss2/4.

3. Cantwell and Van Kampen-Breit, "Librarians and the Institutional Review Board (IRB)."

4. Judith G. Robinson and Jessica Lipscomb Gehle, "Medical Research and the Institutional Review Board: The Librarian's Role in Human Subject Testing," *Reference Services Review* 33, no. 1 (2005): 20–24.

BIBLIOGRAPHY

Cantwell, Laureen P., and Doris Van Kampen-Breit. "Librarians and the Institutional Review Board (IRB): Relationships Matter." *Collaborative Librarianship* 7, no. 2 (2015): 66–78. https://digitalcommons.du.edu/collaborativelibrarianship/vol7/iss2/4.

Steelman, Susan C., and Sheila M. Thomas. "Academic Health Sciences Librarians' Contributions to Institutional Animal Care and Use Committees." *Journal of the Medical Library Association* 102, no. 3 (July 2014): 215–19.

Part I

INSTITUTIONAL REVIEW BOARDS

Chapter One

An Introduction to the Institutional Review Board and Its Role in Human Subject Research

Eric D. Albright and Susan M. Harnett

This chapter will describe the charge of the Institutional Review Board (IRB), its composition and membership duties, the review and monitoring process, and informed consent and human subject protection. It will furthermore detail the role of librarians as members or consultants and offer suggestions for additional ways for librarians to become involved with institutional research. For the purposes of this chapter, the IRB is defined by the Food and Drug Administration (FDA) as

> an appropriately constituted group that has been formally designated to review and monitor biomedical research involving human subjects. In accordance with FDA regulations, an IRB has the authority to approve, require modifications in (to secure approval), or disapprove research. This group review serves an important role in the protection of the rights and welfare of human research subjects.
>
> The purpose of IRB review is to assure, both in advance and by periodic review, that appropriate steps are taken to protect the rights and welfare of humans participating as subjects in the research. To accomplish this purpose, IRBs use a group process to review research protocols and related materials (e.g., informed consent documents and investigator brochures) to ensure protection of the rights and welfare of human subjects of research.[1]

The Institutional Review Board (IRB) plays a vital role in human subject research and is federally mandated to review and monitor all federally funded biomedical and behavioral research involving human subjects as participants. The protocol review process serves to safeguard participants in clinical or behavioral research so they have an understanding and awareness of what participation entails, so that they may make an informed decision. It also ensures the research is beneficial to and progresses science while causing the least possible harm to human subjects. Librarians can enhance the work of researchers and investigators as members of the IRB and in complementary roles as research support.

COMPOSITION OF THE COMMITTEE AND MEMBERSHIP

The IRB's composition and membership is codified in 45 CFR 46 §107, which describes the composition of the committee as "at least five members, with varying backgrounds to promote complete and adequate review of research activities commonly conducted by the institution."[2]

IRB members should possess sufficient expertise to review research activities and apply federal regulations and laws, as well as uphold professional scientific conduct and practice. The IRB must include at least one scientific member, who is sufficiently experienced and expert enough to speak to the merit and sound scientific premise of the research. The board must also include at least one nonscientist and a non-affiliate member who cannot be associated with or the relative of anyone associated with the institution. The board cannot be populated with individuals from one department; it must be diverse in terms of expertise, gender, and cultural background. IRB members should, ideally, be representative of varying constituencies and demonstrate "sensitivity to such issues as community attitudes, to promote respect for its advice and counsel in safeguarding the rights and welfare of human subjects."[3] As human subject research may involve vulnerable populations—such as pregnant women, children (individuals under age eighteen), prisoners, and the mentally or physically disabled—the board should consider individuals who are knowledgeable of or advocates for these populations as part of their membership composition decisions. These individuals may serve as members or may act as consultants to the board, and they may include bioethicists, clergy, social workers, and others able to act as representatives of their respective communities. The institution must select an institutional official (IO) who acts on behalf of the institution and ensures that research conducted therein complies with all federal regulations. Each IRB will also select a chair of the committee, who will conduct meetings and liaise with the IRB administrative staff, investigators, and the IO. Though the IO and chair work closely, and in conjunction with other research support departments, the chair, typically a volunteer position, does not report directly to the IO. He or she serves to represent the IRB's mandate to evaluate and protect human subjects in research while the IO's primary concern is to protect the institution by ensuring compliance.[4]

Librarians may participate as nonscientist members, or as scientist members provided they have sufficient expertise in a scientific field. Librarians may also serve as non-affiliated or community members if they are on an IRB at an institution other than their home institution, which would include service on a commercial IRB like the WIRB-Copernicus Group or Advarra. Additionally, librarian members may serve as voting or nonvoting members and may be assigned to review protocols. It is each individual board's decision how best to utilize its members' expertise. Librarians may also serve the IRB as consultants or ex-officio members. Librarians with sufficient expertise may also serve as chairs of the committee. A librarian's primary role may be to support the board, investigators, and the institutional research effort by providing timely and relevant information services, examples of which can be found throughout this book. Librarians may also assist in reviewing informed consent documents for clarity, language, and readability, and research policy issues and laws. Hospital librarians, in particular, may lend their expertise and experience with clinical procedures and processes to inform the board during the review process.[5]

TRAINING AND CERTIFICATION

Most IRB members receive at least initial background and basic ethical training and certification through the Collaborative Institutional Training Initiative, (CITI). CITI provides online educational and certification resources on a variety of topics related to the ethical conduct of research and the history of research ethics. Each institution can select modules for their IRB members. When those are completed, CITI sends a confirmation to the host institution and to the individual member. IRB members who conduct research as investigators may also be required to take additional modules relevant to that role. One requirement of protocol review is to confirm that all investigators and study personnel have taken the appropriate CITI training, as required by the institution or grant funder.

It is also recommended and helpful for IRB members to have a foundational understanding of the history and ethics of conducting human subject research. Members should review seminal documents such as the Belmont Report, the Nuremberg Code, and the Declaration of Helsinki. Members should also familiarize themselves with the relevant laws governing human subject research: 45 CFR 46 (Protection of Human Subjects); 21 CFR 50 (Protection of Human Subjects, Food and Drug Administration); and 21 CFR 56 (Institutional Review Boards, Food and Drug Administration) are the most frequently cited federal regulations. For a more complete legislative history of the IRB, see chapter 2 (Wendy Highby). Members should also be aware of any state or local regulations governing human subject research as well as any institutional policies.

For professional development opportunities, Public Responsibility in Medicine and Research (PRIM&R) and Massachusetts Society for Medical Research (MSMR) are two independent organizations that sponsor trainings and provide resources for IRB members. Resources may also be found on the National Institutes of Health (NIH) Office of Extramural Research (OER) and the Health and Human Services (HHS) Office of Human Research Protection (OHRP) websites. Additional important offices and sites are noted in subsequent chapters as well as in the list of resources found at the end of this book.

THE APPROVAL PROCESS

Investigators must have IRB approval before enrolling any participants, and researchers cannot collect any data without approval from the IRB. The IRB may require any data collected before approval be discarded; collecting data before IRB approval may jeopardize the investigator's research and awards. Investigators must provide the IRB a complete research protocol and all necessary documents, such as the informed consent or supporting documentation (e.g., survey instruments, recruitment tools, etc.). While it is not mandatory to provide a literature search as part of the application process, requirements vary from institution to institution, with some institutions requiring a consultation with the librarian for all new protocols. As described in chapter 3 (Shields and Sarino), a comprehensive literature search or consultation with a librarian may be considered a best research practice. The 2001 death of a study participant in a trial at Johns Hopkins, and the determination that a more complete literature search would have prevented her death, served to underscore the importance of a librarian's expertise as an information expert.[6]

In addition to the IO, an IRB may also have a dedicated office for administrative tasks. Staff will perform an initial review of all submissions and supporting documentation for completeness and accuracy; and they may make recommendations as to the category in which the protocol should be reviewed and assign them to the appropriate reviewer. Research protocols are assigned to one of three categories: *exempt, expedited,* or *full review.* Each category has a different process for approval. The categories are as follows:

Exempt studies[7] do not need an IRB review; these are studies that have no or very minimal risk to human subjects. An example of an exempt study might be a chart review, or a study using de-identified tissue bank samples. This determination is usually made in the office and the study is given an exception by an IRB administrator.

Expedited reviews[8] are performed by the IRB committee chair and one other member of the committee. An expedited review is a complete review of the study by only two committee members; this type of study does not meet the criteria for needing review by the full committee. All of the studies at this level are no more than minimal risk and may include such studies as a questionnaire or survey. Minimal risk may include the collection and protection of personal or privileged information that may identify study participants. Librarians with expertise in assessment or data management may act as consultants for researchers developing an expedited study.

Full committee reviews[9] involve the entire committee and are studies that include more than minimal risk to the research subject or involve protected groups of people as part of the research. Protected groups of people include children, prisoners, pregnant women, and handicapped or mentally impaired people. Risk can include concerns other than bodily or psychological injury; for example, the board should consider risks to data protection and participant privacy. Protocols must be voted on, and the full committee must maintain a quorum of board members to conduct a vote. Committee members may require investigators to provide additional information, documentation, or corrections, and may approve the research conditionally pending recommended and minor corrections. The committee may also table or reject the research, citing major corrections, lack of scientific merit, or unreasonable risk to subjects. The IRB can suggest a proposal be moved to a different category if necessary, based on the category for which they slate their proposal and the risks, populations, and other variables involved in the study. The IRB does not make this change but rather communicates its belief that the study should be proposed for a different review category. The committee will recommend any changes required to move forward with the proposal; and informs the primary investigator (PI). An example might be a study submitted as an expedited review with minimal risk that is determined to involve a protected class of subjects. The board may opt to move this study to full committee and recommend additional protections; if the PI does not follow this recommendation, it is unlikely that the proposal will be accepted by the IRB and thus it is unlikely to be approved research. An investigator may be invited to the full committee meeting to explain his or her research and answer any questions from the board.

When a study comes before the full committee, at least one scientific member (one qualified to evaluate the study) reviews the protocol to assess the risks and how they are being minimized in the research. A secondary reviewer may also be assigned to review the protocol. The reviewer(s) must also consider whether the potential risk to the participants outweighs the benefit to the individual or to the advancement of science. There are a few basic strategies for reviewers to consider when looking at a study. First, the study should adhere to institutional and governmental policies and the principles outlined in the Belmont Report. The study should be well designed, with appropriate scientific and statistical validity. The protocol must detail the steps taken to analyze any data collected from the study and how it will be used. The study group should be sufficient to ensure that researchers can expect to answer the study question. This means that human participation will not be wasted on meaningless or duplicated research. The research procedures should address an equitable selection of research subjects to help ensure the applicability of the findings and must clearly describe inclusion and exclusion criteria for participants. If participants are to be compensated, the proposed compensation must not be such that it might be considered coercive. Other forms of compensation might be considered, such as reimbursement for travel.[10] It is generally safer if the research makes use of procedures that are already being used for patients; this helps ensure the procedures are performed consistently and with proper resources. The study should be continuously monitored for adverse events or other issues that may impact the safety of participants; all adverse events should be documented and reported to the board and/or the appropriate agency as required. The researcher should always consider the safety of vulnerable populations and make sure that those populations are adequately protected. A final significant factor to address is the safety of information regarding the participant to make sure that the data is appropriately stored and secured. Improper protection of privileged health information may be considered a non-compliance. Once initial discussions have taken place, and all members have the opportunity to address concerns, the reviewers may move to approve the protocol as is or recommend minor or major revisions.[11] Investigators must have full IRB approval before initiating recruitment of subjects.

The IRB may also be asked to approve, on an emergency basis, investigational drugs or medical devices for compassionate use. A petition for compassionate use must meet standards for emergency use set by the FDA. The petition must demonstrate a life-threatening situation for which no other acceptable treatment exists; and the situation must be urgent enough such that convening a full board in a timely fashion might be difficult. Compassionate use petitions may involve a single patient, and the study investigator should submit the petition and documentation. Approvals must be reported to the FDA within five days of emergency use.[12]

Informed Consent

The reviewer(s) will also evaluate any informed consent documents. These documents must be provided to each study participant to ensure a thorough understanding of the research and any risks, benefits, costs, and expectations as a result of their participation. The subject, after reviewing consent documents, is free to participate or decline participation. Participants should also be able to cease participation in the study at any time. Declining participation will not affect a patient's treatment plan or care. Consent documents may also be developed for legal representatives of patients who cannot consent due to age, or psychological, intellectual, or physical disability. The informed consent document should be written at a level easily understood by its intended audience, including a level appropriate for child participants. The form should not include language that is misleading or intended to protect the researchers. The standard recommendation is no higher than an eighth-grade reading level; however, a readability assessment conducted in 2003 indicated that informed consent documents had an average readability of tenth grade.[13] Librarians, especially those with an interest in and experience with health literacy, can help ensure that all documents are free of jargon, acronyms, discipline-specific terminology, and coercive statements. The librarian's role on the IRB is often to review the informed consent form (ICF), either as a member of the committee or as a consultant to support both the reviewer and the researcher. The basics of informed consent as outlined in 45 CFR 46 §116 are as follows:

• "Informed consent must begin with a concise and focused presentation of the key information that is most likely to assist a prospective subject or legally authorized representative in understanding the reasons why one might or might not want to participate in the research. This part of the informed consent must be organized and presented in a way that facilitates comprehension."[14]
• The information needs to be detailed enough to allow for an informed decision.
• The form cannot in anyway exempt the researchers from being charged with harm.
• It needs to include a statement that the study involves research and must fully explain the purpose of the research.
• It should describe how long the research will continue and describe all steps for the participant.
• It needs to include a description of any reasonably foreseeable risks or harms to the subject.
• The form should list any potential benefits to the subject from the research.
• The information must disclose any alternative procedures or courses of treatment, if any, that might be advantageous to the subject. This is the "gold" standard or traditional treatment that the subject would receive if they were not in the study.

- Researchers must describe the steps taken to protect the privacy of the person involved in the research.
- If the research involves more than minimal risk, then it must describe how to seek any compensation for injury. It needs to include contact information for the research team and for the IRB as an outside party.
- Finally, the ICF must stress that participation is voluntary, and that the participant can withdraw at any time without any penalty or loss of benefits.

Since the IRB is composed of people from different research and non-research fields, it is important that the protocol be presented in clear language. Many IRBs require a plain language summary of the research as part of the submission. The ICF may be discussed as part of the submission packet at a full committee meeting; clarifications may be suggested before the final IRB approval is given.

Post Approval Monitoring, Adverse Event Reporting, and Non-Compliances

Approval for the research may need to be renewed on a regular basis, if ongoing, and reports given annually to the IRB office. Any amendments to the protocol, procedures, and so forth, must also go through the IRB before they are implemented, so that the IRB can approve any study changes.[15] In general, studies whose initial review was full board will go before the full board again to approve any renewal. Research studies seeking FDA approval for a drug or device need to follow the standard IRB rules. Multinational studies submitted to other countries for approval usually need to comply with the Good Clinical Practice (GCP) guidelines.[16] The GCP guidelines are nearly identical to the IRB guidelines and intended to provide ethical research standards for human subjects on an international scale. Studies involving researchers at multiple institutions may need to go through the IRB at each institution involved, or they may be extended approval based on approval by another institution's IRB—this will depend upon the practices of the researchers' individual institutions. Researchers should discuss a multi-institutional study with the IRB chair or administrative representative in advance to confirm how they should proceed with their proposal. Studies with national recruitment should consider national-level regulations as well as institution-specific regulations for consent, data storage, and more in order to remain in compliance with institutional, local, state or national laws. Protocols involving international recruitment may also be subject to an IRB or similar approval process, according to the laws of the country in which the research will be conducted.

Any adverse events that arise during the research must be reported to the IRB immediately; they will then assess the situation and decide whether the adverse event is related to the study.[17] The IRB may then develop an action plan to remedy the situation. In some cases, the research may to be suspended or halted entirely. The IRB may report such events to the FDA, Department of Health and Human Services (DHHS) or other government agencies, or the study sponsor. The librarian's expertise may be helpful in determining whether similar adverse events have been reported, and what, if any, interventions were undertaken. As an example, an ex-officio librarian performing real-time searches during a full committee meeting was able, based on the report of an unexpected adverse reaction, to locate literature indicating that European studies using that particular dosage of the study medication had in fact noted similar adverse effects, and had banned the medication at that dosage. The IRB then used that information to suggest a lowered dosage of the medication to mitigate future events.

The IRB must also serve to monitor investigators' compliance with the law.[18] The IRB and the institution must develop a written policy concerning non-compliances and how they are handled. Any deviation from the approved protocol, no matter how minor, might be considered a non-compliance. Non-compliances may be self-reported, as in a researcher realizing that a study medication was not stored as outlined in the protocol; or they may result from anonymous, or whistleblower, complaints. The IRB must take any report of non-compliance seriously and must determine the appropriate course of action. Often, a subcommittee comprised of board members may be appointed to investigate the non-compliance. Reparative actions may range from requiring investigators to retrain staff on proper procedures to shutting down an investigators' protocol and possibly his or her total research. Non-compliances may be reportable to the IO, and to other agencies such as OHRP or FDA, depending on severity. Therefore, it is imperative that investigators and the IRB work closely together to avoid these situations and to produce the highest quality research.

MAKING THE CASE TO SERVE

For many health science librarians, the opportunity to serve as an IRB member may be an infrequent opportunity. There is little in the published literature that documents the experience of librarians serving on the IRB,[19] though librarians are recognizably valuable assets to institutions or hospitals with robust research programs; and many librarians are familiar with the IRB submission and approval process as a result of their own research. Library administrators and librarians can advocate for inclusion on the board, citing the advanced skills of health science librarians in data management, grant writing, assessment, research, bioinformatics, and reproducibility. Librarians can also use inclusion on the board as a mechanism and opportunity for expanding services to the research community, including acting as a liaison between investigators and other research offices. Service on or to the IRB is a high impact, high visibility activity that establishes both the librarian and library as research partners, rather than information gatekeepers and knowledge warehouses.

NOTES

1. Office of the Commisioner, "IRB-FAQs," U.S. Food and Drug Administration (FDA), accessed May 11, 2021, http://www.fda.gov/regulatory-information/search-fda-guidance-documents/institutional-review-boards-frequently-asked-questions.

2. Office for Human Research Protections (OHRP), "2018 Requirements (2018 Common Rule)," U.S. Department of Health and Human Services, accessed March 10, 2021, https://www.hhs.gov/ohrp/regulations-and-policy/regulations/45-cfr-46/revised-common-rule-regulatory-text/index.html#46.107.

3. Office for Human Research Protections (OHRP), "2018 Requirements (2018 Common Rule)."

4. Ernest D. Prentice, Sally L. Mann, and Bruce G. Gordon, "Administrative Reporting Structure for the IRB," in *Institutional Review Board: Management and Function,* ed. Robert J. Amdur and Elizabeth A. Bankert (Sudbury, MA: Jones & Bartlett Publishers, 2002): 35–36.

5. Katherine Stemmer Frumento and Judith Keating, "The Role of the Hospital Librarian on an Institutional Review Board," *Journal of Hospital Librarianship* 7, no. 4 (2007): 113–20.

6. Judith G. Robinson and Jessica Lipscomb Gehle, "Medical Research and the Institutional Review Board: The Librarian's Role in Human Subject Testing," *Reference Services Review* 33, no. 1 (2005): 20–24.

7. Ernest D. Prentice and Gwen S. F. Oki, "Exempt from IRB Review," in *Institutional Review Board: Management and Function*, ed. Robert J. Amdur and Elizabeth A. Bankert (Sudbury, MA: Jones & Bartlett Publishers, 2002): 111–13.

8. Gwen S. F. Oki and John Zaia, "Expedited IRB Review," in *Institutional Review Board: Management and Function*, ed. Robert J. Amdur and Elizabeth A. Bankert (Sudbury, MA: Jones & Bartlett Publishers, 2002): 114–17.

9. Susan Z. Kornetsky, "Overview of Initial Protocol Review," in *Institutional Review Board: Management and Function*, ed. Robert J. Amdur and Elizabeth A. Bankert (Sudbury, MA: Jones & Bartlett Publishers, 2002): 143–151.

10. Kornetsky, "Overview of Initial Protocol Review."

11. Kornetsky, "Overview of Initial Protocol Review."

12. Sherry Bye, "Revisions of an Approved Protocol," in *Institutional Review Board: Management and Function*, ed. Robert J. Amdur and Elizabeth A. Bankert (Sudbury, MA: Jones & Bartlett Publishers, 2002): 289–92.

13. Elizabeth A. Bankert and Robert J. Amdur, "'Compassionate Use' and 'Emergency Exemption' from IRB Approval," in *Institutional Review Board: Management and Function*, ed. Robert J. Amdur and Elizabeth A. Bankert (Sudbury, MA: Jones & Bartlett Publishers, 2002): 129–31.

14. Michael K. Paasche-Orlow, Holly A. Taylor, and Frederick Brancati, "Readability Standards for Informed-Consent Forms as Compared with Actual Readability," *New England Journal of Medicine* 348, no. 8 (2003): 721–26.

15. Office for Human Research Protections (OHRP), "2018 Requirements (2018 Common Rule)."

16. Anushya Vijayananthan and Ouzreiah Nawawi, "The Importance of Good Clinical Practice Guidelines and Its Role in Clinical Trials," *Biomedical Imaging and Intervention Journal* 4, no. 1 (2008), e5. https://doi.org/10.2349/biij.4.1.e5.

17. Karen A. Hansen. "Protocol Renewal," in *Institutional Review Board: Management and Function*, ed. Robert J. Amdur and Elizabeth A. Bankert (Sudbury, MA: Jones & Bartlett Publishers, 2002), 293–96.

18. Ernest D. Prentice, Kevin J. Epperson, Christopher J. Kratochvil, and Bruce G. Gordon. "IRB Review of Adverse Events," in *Institutional Review Board: Management and Function*, ed. Robert J. Amdur and Elizabeth A. Bankert (Sudbury, MA: Jones & Bartlett Publishers, 2002), 297–302.

19. Laureen Cantwell and Doris Van Kampen-Breit, "Librarians and the Institutional Review Board (IRB): Relationships Matter," *Collaborative Librarianship* 7, no. 2 (2015), 66–78.

BIBLIOGRAPHY

Bankert, Elizabeth A., and Robert J. Amdur. "'Compassionate Use' and 'Emergency Exemption' from IRB Approval.'" In *Institutional Review Board: Management and Function*, ed. Robert J. Amdur and Elizabeth A. Bankert (Sudbury, MA: Jones & Bartlett Publishers, 2002), 129–31.

Bye, Sherry. "Revisions of an Approved Protocol." In *Institutional Review Board: Management and Function*, ed. Robert J. Amdur and Elizabeth A. Bankert (Sudbury, MA: Jones & Bartlett Publishers, 2002), 289–92.

Cantwell, Laureen, and Doris Van Kampen-Breit. "Librarians and the Institutional Review Board (IRB): Relationships Matter." *Collaborative Librarianship* 7, no. 2 (2015), 66–78.

Commissioner, Office of the. "IRB-FAQs." U.S. Food and Drug Administration. FDA. Accessed May 11, 2021. http://www.fda.gov/regulatory-information/search -fda-guidance-documents/institutional-review-boards-frequently-asked-questions.

Frumento, Katherine Stemmer, and Judith Keating. "The Role of the Hospital Librarian on an Institutional Review Board." *Journal of Hospital Librarianship* 7, no. 4 (2007), 113–20.

Hansen, Karen A. "Protocol Renewal." In *Institutional Review Board: Management and Function*, ed. Robert J. Amdur and Elizabeth A. Bankert (Sudbury, MA: Jones & Bartlett Publishers, 2002), 293– 96.

Kornetsky, Susan Z. "Overview of Initial Protocol Review." In *Institutional Review Board: Management and Function*, ed. Robert J. Amdur and Elizabeth A. Bankert (Sudbury, MA: Jones & Bartlett Publishers, 2002), 143–51.

Office for Human Research Protections (OHRP). "2018 Requirements (2018 Common Rule)." HHS.gov. US Department of Health and Human Services, accessed March 10, 2021. https://www.hhs.gov/ohrp/regulations-and-policy/regulations/45-cfr-46/revised-common-rule-regulatory-text/index.html#46.107.

Office for Human Research Protections (OHRP). "2018 Requirements (2018 Common Rule)." HHS.gov. U.S. Department of Health and Human Services, accessed March 10, 2021. http://www.hhs.gov/ohrp/regulations-and-policy/regulations/45-cfr-46/revised-common-rule-regulatory-text/index.html#46.116.

Oki, Gwen S. F., and John Zaia. "Expedited IRB Review," in *Institutional Review Board: Management and Function*, ed. Robert J. Amdur and Elizabeth A. Bankert (Sudbury, MA: Jones & Bartlett Publishers, 2002), 114–17.

Paasche-Orlow, Michael K., Holly A. Taylor, and Frederick Brancati. "Readability Standards for Informed-Consent Forms as Compared with Actual Readability." *New England Journal of Medicine* 348, no. 8 (2003): 721–26.

Prentice, Ernest D., Sally L. Mann, and Bruce G. Gordon. "Administrative Reporting Structure for the IRB," in *Institutional Review Board: Management and Function,* ed. Robert J. Amdur and Elizabeth A. Bankert (Sudbury, MA: Jones & Bartlett Publishers, 2002), 35–36.

Prentice, Ernest D., Kevin J. Epperson, Christopher J. Kratochvil, and Bruce G. Gordon. "IRB Review of Adverse Events," in *Institutional Review Board: Management and Function*, ed. Robert J. Amdur and Elizabeth A. Bankert (Sudbury, MA: Jones & Bartlett Publishers, 2002), 297–302.

Prentice, Ernest D., and Gwen S. F. Oki. "Exempt from IRB Review," in *Institutional Review Board: Management and Function*, ed. Robert J. Amdur and Elizabeth A. Bankert (Sudbury, MA: Jones & Bartlett Publishers, 2002), 111–13.

Robinson, Judith G., and Jessica Lipscomb Gehle. "Medical Research and the Institutional Review Board: The Librarian's Role in Human Subject Testing." *Reference Services Review* 33, no.1 (2005), 20–24.

Vijayananthan, Anushya, and Ouzreiah Nawawi. "The Importance of Good Clinical Practice Guidelines and Its Role in Clinical Trials." *Biomedical Imaging and Intervention Journal* 4, no. 1 (2008), e5. https://doi.org/10.2349/biij.4.1.e5.

Chapter Two

The Protection of Research Participants in a Political Context

An Institutional Review Board History for Librarians

Wendy Highby

Institutional review boards ("IRBs") are federally mandated committees tasked with proactive review of research plans involving human subjects. These groups were originally required only for grant-funded studies of human subjects. But many institutions have entered into agreements ("assurances") with the U.S. government to submit all human research projects to the review process, regardless of funding source.[1] IRBs are decentralized bodies, managed at the local, institutional level. The purpose of their review is to protect the rights of research participants.

MOTIVATION TO MINDFULLY STUDY IRB HISTORY

Why should librarians study the history of IRBs? Perhaps to resolve the preceding paragraph's cliffhanger—to understand why the research participants, the "human subjects," were (and continue to be) in need of protection. What could go wrong with research, what could turn ignoble in the noble pursuit of knowledge creation? As gatekeepers and collectors/curators of information, librarians know about greed firsthand—not the desire for money[2] but greed for knowledge—we'd like to know everything (or at least *where to find* everything). We understand the madness of the scientist; we can empathize with the empirical quest and its insatiable pursuit of reproducible results (not unlike the quixotic attempt to make all knowledge accessible). Scientific research goes awry when this pursuit eclipses the ethical duty not to harm fellow humans.

The federally mandated IRB-based system of review was "born in scandal."[3] It was legislated in response to sociopolitical pressures stemming from investigative journalists' revelations of harm done to human participants in biomedical research projects. Each of these scandals is recognizable in one word: Dachau, Willowbrook, Tuskegee. Each is synonymous with turpitude and when pronounced, sounds the proverbial warning to know this history so you do not repeat it.[4] Likewise, the edifying parts of IRB history provide patterns worth emulating. We can draw inspiration from the aspirational yet imperfect ethical codes and earnest rubrics produced by the tireless work of committees and commissions. Members of these collaborative groups dedicated careers to agency administration, civil service, advocacy, education, adjudication, and political representation. The following names provide a counterpoint to the cautionary ones and evince respect when intoned: Nuremberg, Beecher, Belmont.

It is important to know the nuanced history behind the incendiary headline, the clever meme, or the fleeting "flashbulb" memory.[5] The historically grounded study of research ethics should be part of every librarian's foundational and continuing educational curriculum. That study should be sociopolitical in context, examining any and all imbalances of power. Of necessity, the IRB history provided herein is simplified because of the limitations of chapter length. It may be considered reductive unless augmented with the further reading indicated in the bibliography. Follow the footnotes, link to the full text, delve in. Read and meld minds with the cited scholars and civil servants, transforming them into mentors and teachers. This chapter, a one-dimensional syllabus of "IRB History for the Librarian," will remain flat without the insertion of your intellectual curiosity and the interpolation of your personal and professional experiences. Participate reflectively (self-reflexively), maintaining awareness of your subjectivity. I present here a library-centric, discursive narrative told through my subjective voice, influenced by eleven years of experience on my local IRB.[6] Mindfully I bring my whole self to this page, and I invite you to bring yours. Together as colleagues, let's proceed. Tell me, what has your experience been?

TIMELINE

Because IRBs are decentralized, locally deployed, there is likely an IRB near you, at a college, university, or medical center. We all benefit directly or indirectly from scientific research that advances knowledge (the development of the COVID-19 vaccines the most recent, high-profile example in 2020). In your roles as citizen, educator, and information professional, you are, at the least, indirectly affected by IRBs—and in many cases, directly involved as an investigator, board member, administrator, or participant. A basic knowledge of the IRB process is essential for all citizens. Insert yourself into the IRB history timeline that follows below. Timelines are helpful tools, aiding us in comprehending this bureaucratically complex, multi-decade history. Make the story personally meaningful by positioning yourself and your local IRB timeline (see chapter 12). It can be as streamlined or complex as you like. In mine I emphasize human rights and points of political crises (created by muckraking journalists' exposés). Then I overlay it with the three eras of administrative history identified by Dr. Charles MacKay of the National Institutes of Health (NIH).[7]

The narrative of this timeline begins in 1945 (see figure 2.1). The United States is coming to terms with Nazi wartime atrocities through the Nuremberg tribunal. The United Nations is formed in this increasingly globalized environment. Unfortunately, the tribunal did not signal the end of scandalous research. Violations continued, arguably because of the lack of a formalized requirement for independent ethical review. In the 1950s, construction of the regulatory infrastructure begins, but it lags behind the burgeoning research. The regulatory system increases in bureaucratic complexity in the 1960s and 1970s with the development of agency infrastructure and ethical codes. In that idealistic era, a series of bioethical scandals are revealed by investigative reporters and those revelations spur the adoption of regulatory initiatives, including the 1974 National Research Act. That act engenders the collaborative authorship of the Belmont Report and other reports. This timeline is not exhaustive but merely representative, and its starting point is arbitrary. The story could have begun much earlier. For example, Sparks's NIH timeline begins with the Hippocratic Oath in antiquity, then leaps to the 1938 passage of the Food and Drug Act (NIH 2002).

Proceedings will soon begin at Nuremberg to establish the guilt or innocence of Nazi physicians who inoculated prisoners in concentration camps with various

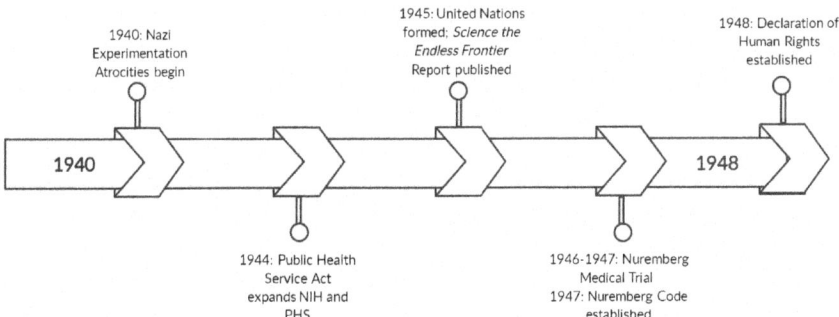

Figure 2.1.   IRB History Timeline, 1940–1949.
L. P. Cantwell

diseases and in other ways shocked public opinion throughout the world. No doubt the defendants will plead that they were engaged in important scientific research for the benefit of society. For this reason, it is important to charge the defendants with genocide . . . that the outrages perpetrated in the name of science in the concentration camps fall within this definition there can be no doubt in the light of the evidence."[8]

1946–1947: Nuremberg: Outrages Perpetrated in the Name of Science and Resistance to Independent Review

The *New York Times* excerpt above foreshadows the purgative aspects of the prosecution of genocide. Nuremberg is the story of coming to terms with our atrocious capacity for evil and appalling lack of empathy for research participants. It is also, arguably, a cautionary tale warning against projection of evil entirely upon others (in this case, Nazi Germany) while deflecting it away from oneself. As we shall learn infra, latter-day historians and bioethicists make the case that Nuremberg's lessons are instructive for scientists of all nations and regimes. Evil cannot be expunged; rather, our awareness of self and others must be heightened, and our ever-present vigilance honed.

The doctors' trial was part of the 1945–1949 series of post–World War II trials sited in Nuremberg (Library of Congress 2021). Also known as the Nuremberg Medical Trial, it was held from 1946 to 1947. Twenty-three scientists and physicians were charged with conducting medical experiments that involved the torture and murder of concentration camp prisoners. Chief Prosecutor General Telford Taylor declared in his opening statement: "This is no mere murder trial. . . . [The court shall] as an agent of the United States and as the voice of humanity, stamp these acts, and the ideas which engendered them, as barbarous and criminal."[9] The Nuremberg medical trial helped post–World War II global society comprehend the sadistic behavior by identifying and punishing it as genocide and developing an ethical code. The zeitgeist after World War II was one of rebuilding, repair, and development of cooperative solutions. The United Nations issued a Declaration of Human Rights in 1948. These actions arose from a shared spirit of international cooperation and a collective hope that, while human beings are capable of horrific evil, we have the power to prevent unethical conduct and atrocities from occurring in the future. Like the Declaration of Human Rights, the Nuremberg Code was also promulgated in 1948, and both promoted the dignity of the individual.[10]

Postwar society attempted to face the worst of the atrocities and to integrate that knowledge into the collective psyche. Many scholars perceive the integration as only partially successful. They make the case that the codified lessons from Nuremberg were perceived to be applicable only to the Nazi scientists and physicians. Author LaChapelle-Henry observes that U.S. researchers could not see how the code applied to their actions. "Because the Code was formed in response to the barbarous acts of the Nazis, many in the U.S. medical and research community avoided the guidelines." Americans distanced themselves from the code, even though its purpose was to "curb exploitation in all human experimentation" everywhere.[11]

Even those directly involved in crafting the code could not help but compartmentalize its applicability. Shuster notes there were three consulting physicians at the trial who helped to formulate the Nuremberg Code: Leopold Alexander, Andrew C. Ivy, and Werner Liebbrandt. Yet these doctors could not bear to apply the strictures to their own research. They held fast to the traditional Hippocratic oath and continued to rely on individual integrity to bolster morality.[12] For example, Dr. Andrew Ivy refused to acknowledge the universality of the Nuremberg Code. He marginalized the document because of its link to atrocious Nazi medical crimes. Ivy considered these acts to be extreme aberrations that would never be performed by Hippocratic physicians. "In Ivy's mind, being a Hippocratic physician was all that was needed to really protect participants in human experimentation."[13]

The newly minted code ceded control to the patient/participant, whereas Ivy remained fixated upon the authority and integrity of the individual doctor. The code, however, had a wider scope. Its ten principles veered toward the rights of participants and away from those of doctors, providing specific patient rights not included in the Hippocratic oath. The Nuremberg Code states that participants have the authority to protect themselves. They possess the right to consent that is voluntary, competent, informed, and fully understood. They also have the right to withdraw from an experiment prior to its conclusion.[14] The code exposed the power imbalance in the traditional doctor–patient relationship: Physicians were guided by the ethical principle of beneficence, "but tended to 'submerge the patient's authority' even when augmented with informed consent."[15] The inherent paternalism in physician decision making compromised the patient's autonomy and diluted the strength of the Nuremberg Code.

LaChapelle-Henry further deconstructs the dynamics in this power imbalance. She highlights the code's endorsement of rights contained in the U.S. Bill of Rights as well as those "firmly rooted in the international human rights movement."[16] She boldly posits that a fundamental shift is necessary to protect against the "scientific thrust of depersonalization . . . that contributed to the egregious torture and murder perpetrated by the Nazi physician-researchers."[17] Her premise: human rights must supersede discovery in the practice of science and medicine; submersion of the patient's authority must cease. Referring to post-Nuremberg scandals, LaChapelle-Henry champions the individual's human rights, observing that "U.S. history has provided evidence time and time again that abuses can occur when scientific inquiry and breakthroughs come at the expense of the individual."[18]

The indifference toward the code, at least on the part of mainstream America, may really have been ignorance. Blacksher remarks that the U.S. public "remained relatively unaware" of the Nazi doctors' atrocities and the trial. The press provided "limited coverage." It was assumed that "research in the United States occurred only with volunteers" and thus most of the press viewed it with approval.[19] Physicians who read the *Journal of the American Medical Association* (*JAMA*) would have been aware of the doctors' trial and the development of the code as a condensed version was published in the December 1946 issue.[20] Blacksher states that this was "the first time formal guidelines for research with humans were widely available to U.S. medical researchers." She also contends that the physicians who did comply with the new standards for clinical research "were a distinct minority."[21]

In summary, the 1947 Nuremberg Code was the first international code of research ethics. It established widely accepted ethical norms for the conduct of research involving human subjects, including voluntary consent of the individual, sound science, minimal risk, and the subject's right to withdraw from the research at any time. But it did not address independent review of research. The Hippocratic system relies on the ethical behavior of the individual and, thus, is harmonious with the individualism that is characteristic of American culture.[22] It was not until 1953, six years later, that the principle of independent review was officially endorsed. And so, while these ethical norms were widely accepted, they were not, it seems, widely applied.

1953–1963 Regulatory Era:
Endorsement of Independent Ethical Review

Dr. Charles MacKay worked in the Office of Extramural Research at the NIH when he published his overview of the history of the IRB in 1995 (see figure

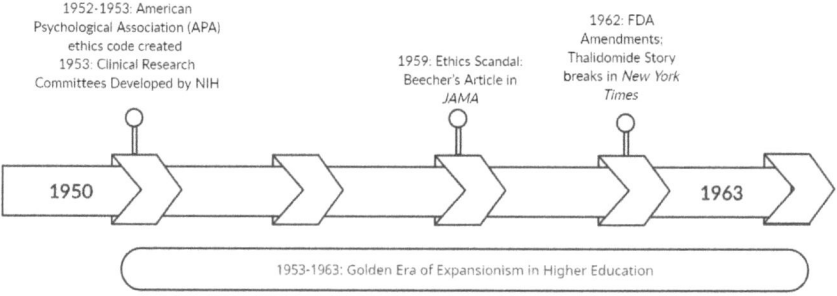

Figure 2.2.   IRB History Timeline, 1950–1963.
L. P. Cantwell

2.2). With several decades of hindsight, he conceived of IRB history as having the following three phases:

1. Endorsement of independent ethical review, circa 1953–1964;
2. Enfranchisement of independent ethical review, circa 1964–1974; and
3. Empowerment of the local IRB, circa 1974–1983.[23]

MacKay points out that the Nuremberg Code, the first international code of research ethics, was "silent on the notion of independent review of the research."[24] After Nuremberg, the next event of note in the IRB chronology is the 1953 opening of the Clinical Center, a research hospital at the NIH. In tandem, Clinical Research Committees were developed by the NIH, and they became the model for contemporary IRBs. Congress used them as a template when it passed the 1974 National Research Act two decades later.[25] Independent review was *endorsed but not mandated* in the contemporaneous NIH document titled "Group Consideration for Clinical Research Procedures Deviating from Accepted Medical Practice or Involving Unusual Hazard." The decision to review was self-imposed, falling upon the shoulders of the physicians doing the research.[26] If the research used patient volunteers or involved "unusual hazard," then written consent was required. Unusual hazard was defined as "jeopardy to the life or relative state of well-being of the research subject," or "jeopardy to the subject's chances of cure of his illness or alleviation of his symptoms."[27] In 1954, a year after the opening of the research hospital, independent review and written informed consent procedures were applied to all intramural research at NIH that involved "normal volunteers."[28]

During the first decade that the Clinical Center was open, from 1953 to roughly 1963, review by groups of experts became the formalized method of governing the intramural research.[29] As for the adoption of independent/peer review on the dispersed institutional level in the 1950s and early 1960s, little has been centrally documented. It "existed in at least some medical schools." Responses received to a questionnaire distributed to medical schools and covering 1961–1962 indicated that "approximately one-third of those responding had committees, and one-quarter either had or were developing procedural documents."[30]

Acceptance of independent review was not the norm in the Cold War era. Most policy makers were hesitant to intrude upon the sacrosanct doctor–patient relationship such that "widespread adoption of ethical principles in the conduct of human studies was slow to develop . . . and until national attention focused on some research scandals in the 1960s, specific human protections in that context seemed unnecessary."[31]

Deference to doctors and squelched scandals aside, concern about legal liability did induce some action during the late 1950s and early 1960s. The identification and development of acceptable standards of care began in this era, as did the documentation of existing practices in reputable research organizations.[32] This evidenced some movement away from the Hippocratic tradition of self-reviewed independence and toward independent review by peers.

On the broader social front, a heightened consciousness, new ideas, and unrest were simmering in the late 1950s and early 1960s, with the United States at the cusp of the transition from the Cold War era to the Civil Rights era. The influence of Nuremberg slowly seeped through the medical community. The National Conference on the Legal Environment of Medical Science held in Chicago in 1958 produced a commentary on the Nuremberg Code. The World Medical Association drafted a Code of Ethics on Human Experimentation, and it was presented at the association's general assembly in Geneva in September 1961.[33]

Though small changes percolated within agencies and learned societies, taking the form of guidelines or studies, there seemed to be little public support or political impetus for regulatory or legislative action until the thalidomide tragedy came to light. The inadequately tested drug was administered to pregnant women and resulted in serious birth defects. Public reaction led to major reform of the system of medical drug testing and approval. The 1962 amendments to the Food, Drug and Cosmetic Act and the 1963–1964 Food and Drug Administration (FDA) regulations enacted and implemented these changes.[34] Also known as the Kefauver-Harris amendments to the Federal Food, Drug, and Cosmetic Act, the changes required the FDA to determine the efficacy of new drugs as well as their safety (P.L. 87-781). The amendments also required the informed consent of participants upon whom the new drugs were tested.[35] This was the first U.S. law to require informed consent.[36]

Dr. William J. Curran, professor of legal medicine at Harvard, witnessed the coalescence of bureaucratic and public desire to regulate participant protections. Curran was a visible public intellectual in the 1950s–1970s, best known for his columns in the *New England Journal of Medicine (NEJM)*.[37] For the journal *Daedalus*, he conducted an in-depth comparative analysis of two federal agencies, the FDA and NIH. He studied the differing approaches each had toward regulation, focusing on the timespan from 1959 to 1968. Curran places the influence of political economy front and center, beginning his article with research funding statistics and commenting on the inevitability of regulatory creep. Curran estimated that U.S. research expenditures for medicine and science grew from $161 million in 1950 to $2.5 billion in 1968. Coincident with this increase in expenditures was a growth in public concern for the human subjects of said research, which would then lead to greater public regulation of the use of humans in clinical medical investigations. Before this influx, there existed "a general skepticism" toward the development of ethical codes, guidelines, and procedures.[38]

Curran highlights the innovative review regimen created by the NIH by contrasting it with the traditional approach of the FDA. The FDA, in order to protect the consuming public, is involved in regulating all phases of research, from design to clinical investigation. It assures that investigators obtain the consent of patients and subjects in clinical trials, and that this program of control is uniform throughout the United States[39] On the other hand, instead of applying a regulatory program uniformly across the nation, the NIH has a system of decentralized, institutional review committees and offers generalized ethical guidelines to protect patients and subjects. Curran says, "the review committee structure is an interesting and imaginative approach to law making . . . [and this] system cannot help but have a profound effect on the medical research community."[40] Curran praises the system's capacity to respect academic freedom and engender creativity.

See Stark (2012), a contemporary scholar, for a more detailed—and less sympathetic—latter-day, revisionist view of the political machinations of the NIH than Curran's above. With the vantage point of more than half a century of hindsight and looking through the lens of a sociologist, Stark perceives the situation chiefly in political terms. He sees the inventions of expert review and of bioethics as a profession as "two parallel stories with one common cause: medical researchers' concern over their legal liability in clinical studies and clinical care." He maintains that both inventions (bioethics and expert review) helped researchers deal with a legal crisis that threatened the funding of their research and their reputations. He concludes that "the legacy of this brilliant political one-upmanship remains with us today."[41] Stark's unsentimental views at worst veer toward cynicism and at best provide motivation for reforming the system. Stark addresses fear, shame, and greed head-on when he openly refers to the legal crisis that endangered researchers' "share of the federal budget" and "reputations." His frank approach reminds readers of the importance of considering the political and economic context when studying the social history of research ethics.

Adding Political Economy of Knowledge Creation to the Timeline

1953–1963 Golden Era of Expansionism in Higher Education

To be thoroughly comprehended at a systemic level, the history of the IRB should be seen in context with the political economy of knowledge creation.[42] Academic or special librarians who work at universities and other research-oriented institutions know the economic circumstances are paramount; grants are the funding fuel that enable the university or organization to competitively function as a nexus of knowledge creation. Librarians, who are intimately familiar with the scholarly communication system and function as gatekeepers of information access, find affinity with the gatekeeping function of IRBs. And as subscribers to resources, monitors of serials inflation, promoters of open access, and consultants to faculty researchers, librarians are keenly aware of the costs, rewards, and potential frustrations inherent in the political economy of knowledge creation and dissemination.

Historian Zachary Schrag connects the dots of the causal relationship between the genesis of IRBs and the push to fund research. He contends that IRBs were created in response to the vast expansion of medical research that followed World War II. Congress, in 1944, passed the Public Health Service Act. This act greatly expanded the NIH and their parent agency, the Public Health Service (PHS). In the decade spanning 1947 to 1957, the NIH's research grant program grew from $4 million to more than $100 million. From 1947 to 1966, the total NIH budget increased from $8 million to over $1 billion.[43]

This development of the bureaucratic infrastructure paralleled and was symbiotic with the unprecedented growth of research universities in the United States. Historians Graham and Diamond name the powerful systemic forces of the era that fiscally impacted American universities and colleges: "the rising tide of economic prosperity, the baby boom, and the revolution in federal science policy." The result was a deluge of postwar federal research funding that drove the research economy of American universities and colleges.[44]

This trajectory had begun in the 1940s. Much of the funding was directed toward biomedical research, which was entering a new era of systemization and standardization. This research was a strategic part of the war effort and post-war planning. At President Roosevelt's behest, Vannevar Bush, head of the Office of Scientific Research and Development, authored a report, *Science: The Endless Frontier*.[45] The report included biomedical research plans and the blueprint for postwar government aid for scientific research planning.[46] It "set the stage for much of the institutional structure that followed as well as how that research would ultimately be regulated."[47] In the following decade, the 1957 Soviet launch of Sputnik, the world's first satellite, shocked the nation and continued to invigorate the competitive attitude of the United States and boosted federal support of scientific and biomedical research.

During the Eisenhower and Kennedy presidencies, 1958–1963, the federal government began to shoulder the economic responsibility for basic research support. In some cases, this included not only grants, but also buildings, equipment, laboratories, and graduate and training programs.[48] At the same time, it was decided that "the research enterprise was to be carried out primarily by the nation's universities as an integral component of graduate education." Between 1953 and 1963, a "golden era of expansionism," expenditures on sponsored research at U.S. universities "grew from $255 million to $1.1 billion, and the federal share of the funding grew from 54 percent in 1953 to 70 percent in 1963."[49]

In their description of the research boom at post–World War II U.S. academic institutions, historians Graham and Diamond highlight the ethos of neoliberal capitalism in higher education:

> Few would question the centrality of knowledge creation to the university's mission, or challenge the importance of research in the knowledge-based economy of the future, or quarrel with the claim that leadership in knowledge creation is crucial to academic and national prosperity.[50]

Riding this tide of expansive "national prosperity," heady with the burgeoning higher education budget, and bamboozled by belief in unsustainable, unlimited growth, what could go wrong? Lurking in the collective unconscious was a miasma—the moral turpitude of atrocities that had been exposed at Nuremberg, and then a cascade of concomitant and subsequent incidents revealed by muckraking journalists would follow. This thread of the narrative has a different trajectory, a tragic and compensatory arc in contrast to the foregoing triumphantly competitive one.

1964–1974: The Era of Exposé and Enfranchisement: Additional Outrages, Further Federal Regulatory Remedies

In this second regulatory era delineated by Dr. Charles MacKay, the cascade of events quickens, seemingly influenced by the politically charged zeitgeist of the 1960s (see figure 2.3). The World Medical Association (WMA) pro-

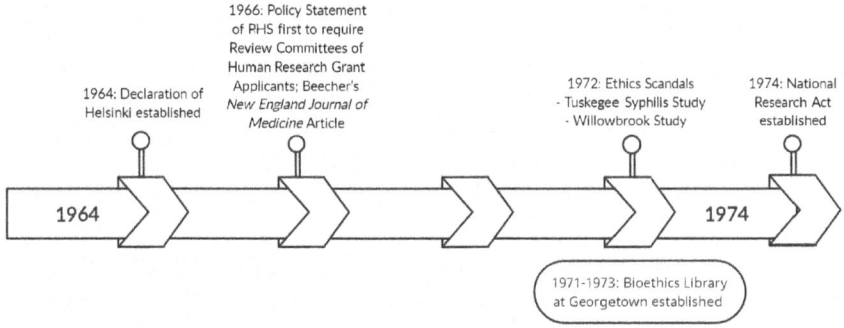

Figure 2.3.   IRB History Timeline, 1964–1974.
L. P. Cantwell

mulgated the Declaration of Helsinki in June of 1964. MacKay sees in the declaration a "clear shift in thinking about human experimentation." Ethical experimentation under the aegis of the declaration now required that "the design and performance of each experimental procedure involving human subjects . . . [be] transmitted to a specially appointed independent committee for consideration, comment, and guidance."[51] The declaration was authored by WMA's Committee on Medical Ethics, chaired by Hugh Clegg, M.D., of the United Kingdom, and then Antonio Spinalli, M.D., of Italy. Work on the document commenced after World War II. Many national medical associations gave feedback. It was adopted in 1964 by the 18th World Medical Assembly in Helsinki, Finland.[52] Distinct from the Nuremberg Code's focus upon research participants' human rights, the Declaration of Helsinki emphasized the duty of physician/investigators to their research subjects.[53]

Also in 1964, a scandal was uncovered by the media—the unethical treatment of research participants at Brooklyn's Jewish Chronic Disease Hospital. Elderly patients were injected with cancer cells without their knowledge. The investigating physician, Dr. Chester Southam of Sloan Kettering Institute, defended his actions to the *New York Times*: "It is not necessary to present [the subject] . . . with what you feel are inconsequential data and it is unethical to ram down his throat information which is detrimental to his condition."[54] The researcher's rationalization is craven and his entrenched and oppositional attitude toward informed consent is at best paternalistic. This case received extensive press coverage. It was the subject of a lawsuit and Dr. Southam was disciplined by the New York State Board of Regents.[55] Another matter that garnered publicity, influenced public opinion, and likely impacted agency policies was the muckraking of anesthesiologist Henry Knowles Beecher. Dr. Beecher was educated at Harvard Medical School and joined its faculty and the staff of the Massachusetts General Hospital in the 1930s. A diligent investigator, he researched the effects of drugs on pain perception in the 1940s and 1950s. He became concerned with determining the efficacy of the deluge of new drugs available. He was aware of many instances of specific research protocols, the ethics of which disturbed him. "He had no trouble accumulating some fifty examples of what he considered investigations of dubious ethicality."[56] He catalogued these incidents and presented them at a medical conference in 1965. Subsequently he submitted the expose for publication. He was rejected by *JAMA*, but later accepted by *NEJM*. His listing included Southam's misdeeds at the Jewish Chronic Disease Hospital. Beecher's presentation caused a stir, was well publicized and may have helped to further the ongoing efforts to tie grant funding to protocol review requirements.[57] While Beecher blew the whistle at the conference and worked with editors to prepare his exposé for publication, the NIH worked to extend the independent review requirement to Public Health Service–funded research.

Charles R. McCarthy, former liaison to the Belmont Commission and director of the Office for Protection from Research Risks (OPRR) at NIH, describes the campaign waged in 1965 by the NIH director James A. Shannon. Shannon introduced a resolution recommending to the National Advisory Health Council (NAHC) that protocol designs of research involving human subjects be submitted to a committee for their independent evaluation of the risks and benefits and "to assure maximum protection for the rights and welfare of those enrolled in each study." McCarthy describes what a sea change this resolution was

> by calling for review by impartial observers, Shannon was reversing two traditions that had been in place at the NIH for more than 25 years. The first tradition held that the physician in charge of the research is the best qualified person to make the judgment of what is in the best interests of research subjects. The second held that nonscientists, that is, laypersons, were not qualified to pass judgment on the ethical aspects of medical research.[58]

The NAHC deliberated Shannon's suggestion and then approved and promulgated it as Policy and Procedure Order #129 the following year, on February 8, 1966. This issuance was the moment of enfranchisement, per MacKay. Below is the pertinent portion of the policy, verbatim. It required the following:

> [T]he grantee institution will provide prior review of the judgment of the principal investigator or program director by a committee of his institutional associates. This review should assure an independent determination: (1) of the rights and welfare of the individual or individuals involved, (2) of the appropriateness of the methods used to secure informed consent, and (3) of the risks and potential medical benefits of the investigation.[59]

This statement, issued by the surgeon general of the U.S. Public Health Service, was the first federal policy statement requiring research institutions (hospitals and universities) to establish the committees that subsequently came to be known as institutional review boards. These IRBs were required of recipients of Public Health Service grants that supported research involving human subjects.

Meanwhile, Beecher's persistence had paid off. His 1959 article in *JAMA* decrying the lack of informed consent procedures for research participants had been largely ignored. His 1966 article in *NEJM* put the economic aspect of the conundrum front and center, literally on the first page of the article. He fingered the enormous and continuing increase of funds available for research as a factor of transcendent importance. Due to a change of zeitgeist, perhaps,

Beecher's impact in 1966 was greater than in 1959. Awareness of Beecher's exposé motivated a member of Congress to write to the NIH inquiring about possible corrective actions. Beecher's research provided support for a 1965 proposal by NIH director James Shannon to require peer review of research, protect the rights and welfare of participants, and ensure appropriate informed consent.[60]

In February of 1966, possibly in anticipation of Beecher's forthcoming exposé in the June issue of *NEJM*, the U.S. surgeon general requested that hospitals and universities establish review boards by the issuance of the Policy & Procedure Order #129. While there is some disagreement among scholars about whether there was any direct causality of Beecher's action upon the promulgation of the landmark policy,[61] the timing was provident. Beecher provided the damning data and agency leadership provided the prophylactic protocol.

Beecher proved to be a bellwether, as scholars enthusiastically responded to his *NEJM* piece and joined the discussion.[62] Historian David Rothman highlights 1966 as the start of extensive changes that would transform both bioethics and the traditional patient–doctor relationship. He sees these changes in "medicine's moral code" as a collaborative process between patients, lawyers, and ethicists. According to Rothman, Beecher "had joined the ranks of Harriet Beecher Stowe, Upton Sinclair, and Rachel Carson" in pioneering significant, consciousness-raising change.[63] The timeline virtually displayed by NIH's History Office identifies Policy & Procedure Order #129 as the origin point of IRBs, the nexus at which human subject research required independent prior review.[64] I argue that the 1974 National Research Act be considered the point of origin for the IRB, as it is more fully buttressed by the creation of a commission with assigned tasks to provide ethical guidelines that will support the entire system of IRBs. But no matter where you pinpoint the origin, both are critical points on the timeline.

With the 1966 precedent set by the U.S. surgeon general's policy, it was only a matter of time before media coverage of emergent scandals would influence public opinion and prod policy makers and lawmakers into actions that would more widely implement participant protections. Nineteen seventy-two was a watershed year in terms of investigative journalists raking through the muck of research ethics-related missteps. Editor and physician Robert A. Greenwald describes the public fervor:

> [N]ewspapers published reports that NIH-supported scientists were perfusing decapitated fetal heads in ketone metabolism studies. . . . [T]his announcement provoked expressions of public outrage. . . . The demand for action and answers was forcefully communicated to those who . . . provided the funds to pay most of the bills, namely the members of Congress.[65]

Two human experimentation debacles emerged in 1972's media coverage most prominently: Willowbrook and Tuskegee. In the former study (spanning 1956 to 1972), children residing at Willowbrook State School for the Retarded in Staten Island, New York, were injected with a form of hepatitis. For some, their admission to the school was conditional upon their parents consenting to the study. Journalist Geraldo Rivera broke the story of conditions at Willowbrook on ABC television.[66] The Tuskegee, Alabama, research project was reported by Jean Heller of the *New York Times*. In this latter study,

> USPHS physicians followed several hundred African-American men who had syphilis. No treatment was given to these men even after the discovery of penicillin. Begun in 1932, the study was not discontinued until 1973. . . . [T]he Willowbrook study had been reviewed by an ethics committee, and the Tuskegee study apparently had also had such a review, but neither study was stopped until the media reports and subsequent public reactions.[67]

Senator Edward M. Kennedy had become chairman of the Health Subcommittee of the Senate Labor and Public Welfare Committee in 1971. The Tuskegee study became a centerpiece of those proceedings, already in progress.[68]

Passage of the National Research Act: Legislative History and Limited Oversight

If the IRB had a birthday, it would be July 12, 1974, the date that President Richard M. Nixon signed the National Research Act. The timing of the IRB's birth was ironic (see figure 2.4). On August 8, 1974, just a few weeks later, Nixon would resign the U.S. presidency because of the ethical imbroglio of Watergate. The ethically aspirational IRB came into being through the signature of a scandal-ridden executive soon to be hoisted on the petard of his own misdeeds. Into the maelstrom of the Watergate scandal, the federally mandated review board was born. The christening of the IRB is explained by Dr. Robert Levine, professor of medicine and bioethics at Yale: the reviewing body required by a 1971 FDA regulation was called the "institutional review

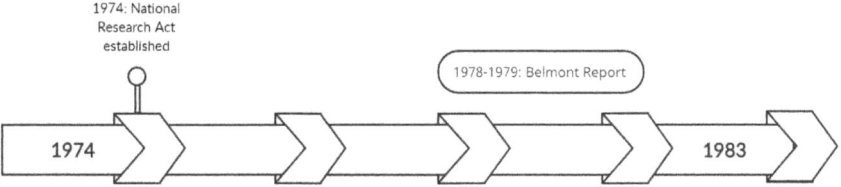

Figure 2.4.   IRB History Timeline, 1974–1983.
L. P. Cantwell

committee"; Health, Education, and Welfare (HEW) proposed regulations in 1973 designating their reviewing body the "organizational review board." The 1974 National Research Act established the name "institutional review board," essentially a portmanteau that Levine calls "a compromise between the two names then extant."[69]

Here is the pith of the legislation, the language of the National Research Act that breathes life into the IRB:

> The Secretary shall by regulation require that each entity which applies for a grant or contract under this Act for any project or program which involves the conduct of biomedical or behavioral research involving human subjects submit in or with its application for such grant or contract assurances satisfactory to the Secretary that it has established (in accordance with regulations which the Secretary shall prescribe) . . . *a board to be known as institutional review board* to review biomedical and behavioral research involving human subjects conducted at or sponsored by such entity in order to protect the rights of the human subjects of such research.[70] (emphasis added)

A few months prior to the July birthdate, on May 30, 1974, the Department of Health Education and Welfare (DHEW) had published NIH policies on human subject protection in the Federal Register (39 105 part 2) noting they were converted to regulatory status and would be codified in the Code of Federal Regulations at 45 CFR 46. This legislation was a watered-down compromise. One bill, proposed by Senator Hubert Humphrey, would have created a salaried, president-appointed board with regulatory and precedent setting authority, and the centralized power to review all federally funded research.

Another bill, proposed by Senator Edward Kennedy, would have created a permanent commission that could issue and enforce regulations and certify local boards. In the diluted version of the Kennedy bill that did eventually pass, the commission became temporary, and oversight was deliberately buried in obscure sub-agencies. This set the precedent for a decentralized pattern of research regulation that was libertarian in the sense that it eschewed "big government" solutions. Sociologist Sarah Babb contends that "centralization was antithetical to the interests of powerful actors. The biomedical research community strongly opposed any potential imposition on their professional autonomy."[71] And the NIH preferred the insulation from controversy that decentralization helped provide. This laissez faire hands-off approach continued until the mid-1990s.[72] With passage and implementation of the act, the federal government had leveraged its grant-funding largesse to catalyze the building of a localized infrastructure of panels tasked with monitoring the morality of biomedical and behavioral research scientists. Babb succinctly describes the

IRB process as "the outsourcing of ethical judgments to local committees with limited government oversight."[73]

1974–1983 Regulatory Era: Empowerment: Belmont and Beyond

With the passage of the National Research Act, we have arrived at the third regulatory era delineated by Dr. Charles MacKay, "Empowerment of the Local IRB." By 1974, twenty-seven years have elapsed since the end of the Nuremberg medical trial in 1947. This gap between revelation of the problem and adoption of a legislative remedy was, arguably, attributable to our failure as a society to integrate the Nuremberg findings about the Holocaust in a reflexive manner. As documented supra, scholars have identified various reasons for this failure: an initial lack of awareness of the general public, disidentification with the extreme cruelty of the Nazi experimenters, and researchers' resistance to the concept of independent ethical review (particularly physicians who were loath to cede their power to independent reviewers and patients/participants).[74]

Forward-thinking policy makers and leaders in the NIH required that the review process be implemented and applied to intra-organizational research projects as much as possible and also externally through the conditional award of grant funds. But eventually, with the raised consciousness of the sixties and seventies (analogous to today's "woke" state of mind), progressive, pro-regulatory political pressure increased from constituents, and legislation was passed. With the passage of the 1974 National Research Act (the "Act"), the nationwide system of IRBs and the federal agencies overseeing the system were given the go-ahead to proceed. Much work lay ahead; it would take four years to begin to gather, cull, and distill the testimonies, public comments, and ethics-related scholarship necessary to create the documentation that would guide local IRBs.

The Act established a committee to be known as the National Commission for the Protection of Human Subjects of Biomedical and Behavioral Research ("the Commission"). The composition of the group was specified as follows:

> (b) (1) The Commission shall be composed of eleven members appointed by the Secretary of Health, Education, and Welfare (hereinafter in this title referred to as the "Secretary"). The Secretary shall select members of the Commission from individuals distinguished in the fields of medicine, law, ethics, theology, the biological, physical, behavioral and social sciences, philosophy, humanities, health administration, government, and public affairs; but five (and not more than five) of the members of the Commission shall be individuals who are or who have been engaged in biomedical or behavioral research involving human subjects.[75]

Many of the appointees were chosen because of their expertise and specialization in the study of the vulnerable populations that would be addressed in the special reports to be written. The appointees would construct the ethical/intellectual infrastructure. Sufficient resources were dedicated to support their endeavor. They were well remunerated (paid at the rate for federal grade GS-18) (Section 201(b)(2)(A) of the Act) and had the assistance of support staff and consultants. Out of the eleven commissioners appointed, five were researchers. Of these, three were physicians and two were psychologists. Among the remaining six commissioners were three lawyers, two ethicists, and a social worker.

The experiences and views of the Commission members, staff, and consultants have been captured by the recording of oral history interviews. These are on display in a virtual exhibit on the website of the Bioethics Research Library at the Kennedy Institute of Ethics at Georgetown University. These digital artifacts provide a unique window into the Commission's process. From the positive nature of the interviews, MacKay's assessment that the Commission was part of an "era of empowerment" seems on point. The process appears to have been well shepherded and well planned. In reading the transcripts of the Commissioners, it is apparent that in order to feel empowered to accomplish their mission, they needed a harbor where they could safely engage in study, debate, critical thinking, and writing, far from the highly politicized environment of the U.S. Congressional floor. The interview of Belmont Commissioner Dr. Duane Alexander describes the intense atmosphere of the Congressional environs. They were grappling with tough issues such as the Tuskegee syphilis study, psychosurgery, research with prisoners and with people who had mental impairments, and fetal research (OHRP Alexander 2004). The construction of a carefully considered philosophical/ethical base was necessary to prepare for the full implementation of the National Research Act. Stakeholders also needed guidelines concerning the ethical treatment of vulnerable populations. The Commissioners served as a panel of experts conducting a careful and deliberative study, and thus, undesired outcomes like reactive legislation and/or hasty implementation were avoided. The Commissioners could deliberate publicly, but without political pressures impinging upon them.[76]

Alexander explains that this outsourcing was also helpful to the DHEW, which was stressed by the high visibility of the Congressional debate and proceeding "very hastily under the pressure of these things happening on the floor of the Congress." By having the DHEW interact with the Commission instead of Congresspersons, some of the pressure was relieved and the regulatory process could be developed in "a more orderly fashion." He concludes that the formation of the Commission was "a very, very wise move."[77] Because the Commission deliberated in public hearings, they "enjoyed support of the public who were given an opportunity to testify at nearly every meeting." They also had good relations with the press, who were also allowed to be present for all of their deliberations.[78] The inclusive process appears to have been empowering for all concerned: the Commissioners, the media, and the public.

The Commission met two days per month for four years, commencing in December 1974 and ending in September 1978. Safely distanced from the fishbowl of the Congressional floor, the group was congenial; they bonded and their workplace became a socially nurturing space, according to Commission staffer and behavioral scientist Miriam Kelty. She describes the camaraderie of the process:

> Well, there were two things that stand out. One was process and one was product. The process thing was that the staff and the commissioners and the people that we worked with worked very closely and very well together. This day, 30 years later, I still feel a kinship with most of those people and still see a number of the people.[79]

These convivial Commissioners had much work to do. The energy surrounding the project was similar to the post–World War II era of rebuilding and repair, in this sense—citizens in the post-Watergate era, while bruised by Nixon's perfidy, could be characterized as an optimistic cohort. A climate of dedication to civil service prevailed and is evidenced in the oral history interviews collected from the Commissioners, and from the staff and consultants working with them.

The Commissioners had a work environment conducive to cooperative critical thinking which increased the likelihood they would have a successful collaborative experience. Having meaningful tasks on their to-do list helped to ensure success as well. Those projects were set forth in the Act as follows:

> 202. (a) (1) (A) The Commission shall (i) conduct a comprehensive investigation and study to identify the basic ethical principles which should underlie the conduct of biomedical and behavioral research involving human subjects, (ii) develop guidelines which should be followed in such research to assure that it is conducted in accordance with such principles, and (iii) make recommendations to the Secretary (I) for such administrative action as may be appropriate to apply such guidelines to biomedical and behavioral research conducted or supported under programs administered by the Secretary, and (II) concerning any other matter pertaining to the protection of human subjects of biomedical and behavioral research.

McCarthy praises the quality of the Commission's reports as "thoroughly re-searched" and "meticulously documented," with sources that "provided depth and credibility seldom associated with reports of federal advisory bodies."[80] This was high praise indeed from a former college professor.

The Commission was productive. Their central handiwork was the issu-ance of basic ethical principles and guidelines in the Belmont Report (the "Report"). The Report identified three tenets: justice, beneficence, and respect for persons. It did not make specific recommendations for adminis-trative action by the secretary of DHEW. Instead, the Commission recom-mended that the Report be adopted in its entirety as a statement of DHEW policy. The subsequent adoption of the Belmont Report represents a rare instance of the federal government formally accepting a moral theory as the foundation for legislation.[81] The Commission prepared ten special reports, ranging from "Research on the Fetus" to a "Special Study on Implications of Advances in Biomedical and Behavioral Research." (See the full text of these and their other reports here: https://www.hhs.gov/ohrp/regulations-and-policy/belmont-report/access-other-reports-by-the-national-commission/index.html; see also "IRB and IACUC Resources and Readings to Know and Explore" list at the end of this book.)

The Bioethics Research Library's virtual exhibit, titled "Belmont Report Anniversary and History," is an edifying record of the Commissioners col-laborative activities and insights. Especially meaningful for librarians is the interview content that indicates the Commissioners (and Commission staff and consultants) valued and understood the significance of the creation of a bioeth-ics library to support not only the Commission's work, but to midwife the gen-esis of a new interdisciplinary field of study. They even recall the few seminal texts first available in that nascent field. This online exhibit commemorates the twenty-fifth anniversary of the publication of the Belmont Report (celebrated in 2004). This trove includes a link to the Belmont Report, an edited version of the twenty-fifth anniversary celebration, an educational video, and an oral his-tory archive of transcripts and embedded videos of interviews with members and staff of the National Commission for the Protection of Human Subjects of Biomedical and Behavioral Research (recorded in 2004). This exhibit brings to the fore and animates the intensive intellectual labor and teamwork required in the crafting of codes, guidance, and protocols. The interviews convey why the task was so intrinsically rewarding—the endeavor was chiefly altruistic. It is remarkable that out of the shame and outrage of Tuskegee, Willowbrook, the Nuremberg doctors' trial, and other scandals arose a collaborative venture that accomplished monumentally good things.

Of particular interest to librarians are the Commission's collection devel-opment efforts. In his interview, Charles McCarthy, the NIH liaison to the

Commission, describes the information gathering phase of the Commission's work:

> I suppose I would say the most important thing the Commission did was the Belmont [Report], but I think perhaps it did a couple of things that go along with that that are equally important. One is simply by gathering all of the information about the ethics of research that it could find in seven different languages and gathering all that background literature, it really in many ways gave birth to, or at least gave . . . an enormous boost to the whole field of bioethics.[82]

McCarthy has an interesting background, having been a priest with a solid academic grounding in philosophy, religion, and political science. He describes the creation of a literature review and a much-needed secular infusion into bioethical philosophy:

> Prior to that time, bioethics was limited to a handful of scholars, and they were almost all moral theologians, so there was virtually no secular writing about the ethics of medicine, or health care, or medical research. The Commission almost single-handedly changed all that. They brought every scholar in the country and many from overseas to testify, they did a literature search into each of the topics under consideration, and then they asked scholars to evaluate all that.[83]

McCarthy's post-priesthood *curriculum vitae* includes a stint at the Virginia Commonwealth University Office of Education and Compliance Oversight. At NIH, he had directed the Office of Protection from Research Risk (OPRR). McCarthy's multi-disciplinary education and subsequent engagement in civil service seem to afford him a deep appreciation of the long-term import and wide-ranging influence of the Belmont project:

> So it was a great shot in the arm to the whole field of bioethics, and people still say today that the Commission saved the teaching of philosophy. . . . [I]t gave an enormous impetus to the whole world of ideas and principles and understandings, not just for research but for philosophy in general and a whole principled approach to guiding human life.[84]

McCarthy also values libraries and, in this excerpt from the interview, he generously champions the Bioethics Research Library (BRL) and its librarians:

> The Kennedy Institute . . . got government grants to create the best library in the world of ethical literature. It's funded by the National Library of Medicine, but it actually is housed at Georgetown University, and I think they now have articles in ethics in about 28 languages, and they have a wonderful set of librarians. All of that came out of the Commission. . . . I just think it's an invaluable contribution to American culture and maybe to world culture.[85]

A history of the Bioethics Research Library is available at this URL: https://bioethics.georgetown.edu/about/history-of-the-bioethics-research-library/. Currently, the BRL contains over one hundred thousand books as well as journals, oral history transcripts and recordings, Bioethics Commission meeting notes and reports, "grey" literature, and unpublished manuscripts. It remains the "world's most comprehensive collection of materials relating to the ethics of health care, biomedical research, biotechnology, and the environment."[86]

Psychologist and Commission staffer Miriam Kelty comments upon the literature that was produced and served to support the pedagogy of bioethics research. She praises the utility of the first textbooks; they provided materials for discussion and teaching and helped to start a new academic discipline regarding the ethics of research.[87] Kelty recalled details about the foundational textbooks:

> There were very few materials available at the time that the commission was established. One was Jay Katz's book. . . . Also, Brad Gray's study had been published quite recently. Other than that, there wasn't a whole lot available. There were court cases and there were other things that were not so easy to come by.[88]

Here, Kelty refers to Jay Katz's 1972 compendium, *Experimentation with Human Beings: The Authority of the Investigator, Subject, Professions, and State in the Human Experimentation Process*, and Bradford Gray's 1975 book, *Human Subjects in Medical Experimentation: A Sociological Study of the Conduct and Regulation of Clinical Research.*

About a year before the release of the Belmont Report, Gray's study of IRBs was published in *Science*. The researchers surveyed sixty-one institutions (a probability sample drawn from the more than 420 institutions with review committees approved by DHEW); it focused upon their reviews spanning July 1974 to June 1975. Almost two-thirds of the reviews were biomedical in nature, and the other third were behavioral. On average, the IRBs reviewed forty-three proposals in a year. Top problems indicated by board members included getting members together for meetings, the need for rapid action to meet deadlines of funding agencies, the lack of precise DHEW guidelines, and the time spent unnecessarily reviewing research with little risk. Consent forms were problematic due to the complexity of sentence structure and terminology used. Study participants were also surveyed, and they offered suggestions about how researchers might improve. The participants expressed the desire that researchers communicate more effectively and that they treat subjects with "greater sensitivity."[89]

As shown by the study results above, the review of research plowed ahead while guidelines struggled to keep pace. The codification of the Belmont

Report and associated procedures and subparts was a protracted process. To recap, the 1974 National Research Act (Title II, Public Law 93-348) was read into the Federal Register and thus the regulatory requirements were codified; all research funded by DHEW was to be reviewed by IRBs. Also in 1974, IRB procedures were established and added to 45 CFR 46, as was Subpart B—Special protections for pregnant women and fetuses. In 1978, Subpart C—Special protections for prisoners—was codified, and in 1983, Subpart D—Special protections for children—was codified. After the Commission disbanded in 1978, a new advisory body was created, The President's Commission for the Study of Ethical Problems in Medicine and Biomedical and Behavioral Research. This new body, known as "the President's Commission," continued on with the work of explicating the ethical foundations of the treatment of human subjects, and produced nine reports between 1980 and 1983.

1983–1995: Counter-Narrative, Common Rule, and Radiation Reportage

MacKay recalls the Belmont era, 1974 through 1983, as the most frenetic, "when the successive statutorily mandated Commissions brought fresh and profound vision that revitalized the system."[90] By comparison, the remainder of the 1980s through the early 1990s were relatively uneventful (see figure 2.5). As director of the OPRR, McCarthy recalls keeping a low profile to stay under the Reagan administration's small government, regulation-cutting radar.[91] Historian Zachary Schrag says little paper trail exists from this era to track what IRBs were doing. He calls this the "age of détente." Similarly, sociologist Sarah Babb refers to it as an era of "approximate compliance." IRBs complied with the regulations approximately; they were guided by informal norms rather than a close reading of the rules. And the OPRR, charged with overseeing compliance, did not have the budget to closely monitor thousands of academic institutions.[92] Schrag provides a marked counter-narrative to the glowing reminiscences of the Commissioners excerpted from the interviews

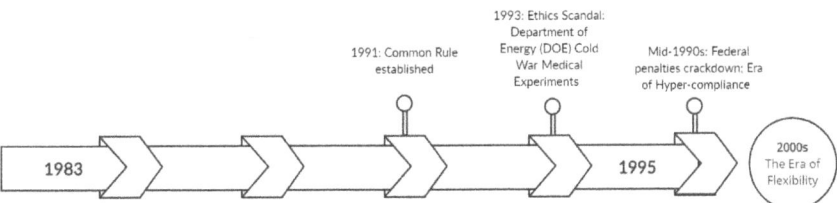

Figure 2.5.   IRB History Timeline, 1983–current.
L. P. Cantwell

above. He believes the "golden days" of research regulation are a myth; "there was just a period of time in the 80s when social scientists were left alone."[93]

While Charles MacKay and Charles McCarthy are agency apologists and provide unabashedly pro-regulatory views in their articles regarding IRB history, Schrag takes a contrarian stance that must be considered and integrated to form a fuller picture of IRB history. Schrag labels Charles McCarthy a "hardened bureaucrat."[94] If the reader prefers not to veer toward sentimental views of IRB history, then the monographs of sociologists Sarah Babb and Laura Stark are recommended. Also read Schrag's chapter 3 "The National Commission,"[95] and his chapter 4 "The Belmont Report," (Schrag 2010, 78–95) in his *Ethical Imperialism* for a critical analysis of the Commission's performance (from the perspective of a significant contingent in the social sciences disciplines).[96]

Schrag ends his monograph with a kind of manifesto, listing six findings in his analysis of the IRB history with regard to its treatment of the social sciences.

1. The "present system of IRB oversight is not based on empirical investigation of ethical abuses committed by social scientists," but instead rests on the record of medical research.
2. Those making policy have not investigated ways to prevent violation of confidentiality, such as finding alternatives to government subpoenas.
3. Social scientists and scholars in the humanities have not been included in the IRB policy-formation process.
4. The regulation of social science was "collateral damage from the attempt to control medical research."
5. The "golden days" of research regulation are a myth; "there was just a period of time in the 80s when social scientists were left alone."
6. "Multi-level federal policy includes: guidance, regulation, and statutory law. The problem with these is that they leave the interpretation of regulations in the hands of officials whose main job is regulating medical research."[97]

Schrag feels that Congress should amend the National Research Act to limit its scope to health-related research. He states, "IRB review of the social sciences and the humanities was founded on ignorance, haste, and disrespect."[98] Schrag's counter narrative is continued on his blog, last updated in July 2018.[99]

All was not rosy on the federal level for "hardened bureaucrat" Charles McCarthy, the head of ORRP and one of the pro-regulation scribes of IRB history. In what seems to be a Kafkaesque scenario, ten years elapsed before the Common Rule was officially adopted and truly held in common by federal agencies. McCarthy ascribes the difficulty of the adoption process to the desire of departments and agencies that regulations be geared to agency-specific programs, not to general standards, and turnover in departments and agencies that necessitated renegotiated agreements with new appointees.[100] Yet McCarthy manages to give it a positive spin, thanking his special assistant Joan Porter for working "tirelessly" on the project. The Common Rule is formally referred to as the Federal Policy for Protection of Human Subjects. The Common Rule includes requirements for (1) assessing compliance; (2) informed consent; (3) IRB membership, function, operations, review of research, and record keeping; and (4) protection for vulnerable research subjects.[101] At culmination, at its point of passage in 1991, the Common Rule had fifteen codifications and sixteen signatories, and it covered eighteen federal agencies. The rule broadened regulated research's scope and provided some cross-departmental standardization.[102] But it did not have a common procedure for interpreting and implementing the regulations and thus, even though departments and agencies were committed to following the Common Rule, the interpretation of regulatory requirements among departments and agencies differed. And this is where things turn Kafkaesque again:

> The Common Rule has four subparts. Subpart A is the only part signed on to by all participating agencies. Subparts B through D address specific additional protections and considerations for research involving fetuses, pregnant women, and human in vitro fertilization (Subpart B), prisoners (Subpart C), and children (Subpart D). Only DHHS and the Department of Education are signatories to Subpart D, and only DHHS adheres to Subparts B and C. EPA has signed on to Subpart A only.[103]

And here is language that only an apologist could have written: "Nonetheless, there are basic concepts contained in the regulations that provide a framework and guidance for federal oversight, even though the specific policies and procedures adopted by a department or agency for implementation might differ."[104]

IRB history, born in scandal, was hit by another bioethical bombshell in 1993. *Albuquerque Tribune* investigative reporter Eileen Welsome told the story of government-funded experiments to determine the effects of radiation exposure on the human body. The research took place from 1944 to 1974 and involved thousands of U.S. citizens; it often resulted in unsafe and even lethal levels of exposure. In many cases, participants were not asked for consent, nor were they properly informed. Many of the investigating doctors and scientists used deception and violated not only their patients' trust but also

flaunted ethical codes (such as the Hippocratic oath, the 1946 AMA guidelines, the Nuremberg Code, and policies of the Atomic Energy Commission
and Department of Defense).[105]

Welsome describes the pervasive and persistent nature of the protective
culture that cloaked anything related to atomic defense:

> The culture bred by the Manhattan Project caused a blanket of secrecy to be
> thrown over everything related to atomic weapons [. . . It was] a reflection
> of the shared attitudes and beliefs of the scientists and bureaucrats who were
> inducted into the weapons program at a time of national urgency and never
> abandoned their belief that nuclear war was imminent.[106]

Welsome lauds the then–energy secretary Hazel O'Leary who undertook
sincere efforts to make all documents public, saying that O'Leary's open
acknowledgement of the extent of the experiment was "a radical departure
for a bureaucracy steeped in secrecy and arrogance."[107] President Clinton
established the Advisory Committee on Human Radiation Experiments to
investigate, and its final report was released in 1995 (DOE). Welsome calls
their findings "disappointing and timid." While the final report contained new
information, its conclusions were weak, and it did not confront the controver-
sial nature of many of the studies.[108]

What can librarians learn from this scandal? According to the old truism,
information is power. To be more accurate, the old saw should be changed to
say that information has the *potential* to be powerful, depending upon how
it is applied. The advisory committee had an estimated six million pages
of documentation made available to them, and they were overwhelmed.[109]
Also, some committee members were hesitant to pronounce judgment upon
individual investigators. In order to deliver a unanimous report, the advisory
committee decided to separate the judgments about "the wrongness of an action" from "the blameworthiness of the person who committed the act. They
separated the experiments from experimenters."[110]

Though not cathartic on a collective level, Welsome's reportage had positive impact on individuals. Elmerine Allen Whitfield, the daughter of one of
the experimental subjects, Elmer Allen, held this attitude toward the new
information: "She often said she couldn't imagine going through life without
knowing what had happened to her father. Although theirs had been a complicated and combative relationship, the knowledge had helped her better
understand him."[111] And library workers were an essential part of this revelation that helped to heal the next generation. Welsome wore out lots of shoe
leather traveling to libraries and tracking down leads over a period of several
years. In her acknowledgements, she thanks archivists and librarians.[112] Some
gatekeepers of information were not so helpful. She encountered resistance at

times, such as the "bureaucratic stonewalling" she received when requesting information from a Los Alamos Laboratory official.[113] The American Libraries Association's ethical code supports unrestricted access to information. This code declares, "We uphold the principles of intellectual freedom and resist all efforts to censor library resources."[114] To be equal to such lofty ideals, librarians must be vigilantly self-reflexive and ask if, when, and how they have withheld or censored, or perpetuated a culture of secrecy? What were the circumstances of the denial of information, and did it involve censorship related to issues regarding "national security"? See the American Library Association's "Issues and Advocacy" webpages for helpful discussion regarding censorship and ethics of the profession.[115]

1995–2011: Federal Crackdown, Concern for Communities, and Change

Zachary Schrag describes the "federal crackdown" or "regulatory chain reaction" that developed in response to the revelations about the radiation experimentation. The new leader of the OPRR, Gary Ellis, took the public criticism to heart. He began "a totally unprecedented flurry of enforcement activity." Starting in 1998, the OPRR "increased its number of onsite investigations and suspended HHS-funded research at eight major universities and hospitals."[116] Sociologist Sarah Babb characterizes this era as one of "hyper-compliance." Gone were the low-profile, laid-back times of approximate compliance—all of this changed in the aftermath of the DOE scandal.

In the midst of the crackdown, there was a change in the federal hierarchy of agencies. The Office for Protection from Research Risks (OPRR), which was created in 1972 and was part of the National Institutes of Health (NIH), was dissolved in 2000. A new agency was created, the Office for Human Research Protections (OHRP); it did not report to NIH but reported directly to the Department of Health and Human Services (DHHS). Some conjectured that this was a political move, to situate the office higher in the hierarchy in order to strengthen the enforcement efforts; others thought it a move to oust Director Ellis, who did not continue with the new organization.[117]

Babb's sociological study covers how IRBs evolved from the 1990s onward. These changes were particularly evident in larger, biomedically oriented research universities. Babb documents spikes in the number of compliance investigations opened[118] and number of enforcement letters sent.[119] Eventually, a flexibility movement took hold and tensions eased. The crackdown subsided in 2003 and the number of enforcement letters declined. This was due in part to the shift in IRB management and membership at many institutions. Originally comprised of faculty volunteers, these boards now included administrators who devoted resources toward federal regulatory interpretation and compliance. She says these changes were prompted by the need to comply with federal audits of the decentralized local committees. Babb identifies two waves of professionalization of IRB work. The first was the hiring of specialized workers to expand IRB staff to handle the increased bureaucratic complexity. The purpose of this first wave was to lessen the risk of non-compliance. The goal of the second wave was to increase efficiency and lower the high cost of compliance. Processes were routinized and many were outsourced to private corporations.[120]

Not only is IRB history born in scandal, it also comes of age in scandal. Bioethical misdeeds are tracked in David Resnick's timeline and, most unfortunately, the scandalous timeline keeps growing. Two particular entries in 2011 will be of interest to librarians. The first is Jeffrey Beall's predatory journals list, described in this thumbnail:

> Jeffrey Beale [*sic*] publishes a list of what he calls "predatory journals." Predatory journals are profit-driven journals that charge high fees for open access publication, promise rapid publication, and have poor (or nonexistent) standards for peer review. Beale [*sic*] later withdraws his list due to pressure from journals.[121]

Beall suffered online harassment from unidentified parties claiming to be colleagues and supporters of open access, and he has been threatened with lawsuits. After shuttering his website in 2017, he commodified his list. It is now produced by the commercial library vendor Cabell's International. The product is in demand because it provides comparative quality indicators of journals and helps academics to more selectively choose a publication venue.[122]

Resnick's timeline can be used to teach the ethics of research. It can be mined for examples to use as case studies. The ethics violations described in the well-reviewed, bestselling book *The Immortal Life of Henrietta Lacks* could serve as a didactic tool. The book helped to raise public consciousness about bioethics and research. Written by journalist Rebecca Skloot, it concerned Henrietta Lacks, an African American woman who provided the tissue for a widely used cell line known as "HeLa." Lacks had been treated for cervical cancer in 1951 at Johns Hopkins Hospital and unfortunately passed away in that same year. Skloot interviewed Lacks's family and learned that researchers had used Lacks's cells without her consent and without compensation. When Skloot discovered this, she decided to share her book royalties with the family. The NIH and the Lacks family reached an agreement in 2013. The family was given control over the data and proper acknowledgment in scientific papers.[123] The National Institute of Environmental Health Sciences

(NIEH) timeline webpage also has links to a "Glossary of Commonly Used Terms in Research Ethics" and a link to Bioethics Resources.

These continuing scandals indicate the need for change. With crisis comes opportunity, an opening to do better. Harkening back to the high hopes of the Belmont Report's creators, Commission staffer and psychologist Kelty explains their intent that the code be lithe enough to accommodate future circumstances. She describes the "intellectual hard work" of critical thinking, the inevitability of change, and the flexibility of the Belmont principles:

> As research changes, then I think the application of Belmont and the rethinking of ethics must be a constant process. So I don't think there is ever a point when you can say: Now we have it all figured out. Belmont is not written as if it's rigid and to be applied, but, rather, it is to guide our thinking. And as long as that thinking is creative and imaginative and concerned for the rights of others, then I don't . . . fear for the research that is to come. But that requires sound and well-educated people to understand the spirit of Belmont and to be willing to do the intellectual hard work to make Belmont realistic instead of simply an icon.[124]

Sometimes the needed changes revolve around our culture's new sensibilities regarding ethnicity, race, gender, indigeneity, disability, and sexual preference. If possible, researchers must alter the way they do research, including the way they approach research ethics. It is time to innovate and experiment (all the while staying within ethical guidelines), and then to document and publish your results. Here is just one example among many. Myra Parker and her colleagues found Belmont's principles lacking when applied to their research with indigenous communities:

> [T]hese principles are based on Western epistemologies . . . which are rooted in Western values and individualism, and thus ignore Indigenous values and privilege the individual over more holistic, community/group perspectives. These gaps in the Belmont principles, therefore, result in the failure to consider community–level risks, and constrains the interpretability and direct relevance to ethical implementation of health research in racially, ethnically, and culturally diverse communities.[125]

But Parker does not see opposition, so much as a lack or gap, a system devoid of community context. Perhaps, she supposes, it would be possible to retain the Belmont baby while throwing out the Eurocentric bathwater:

> [Indigenous] values include the same principles of respect, justice, and beneficence/non–maleficence; however, unlike most Eurocentric individualistic approaches, Indigenous communities extend these principles to the rights and

welfare of communities. Many researchers now apply non–Western thinking to their research methods, but not necessarily to their research.[126]

There are many avenues to follow in pursuing change. Perhaps the above struggle to update ethics to match research methods grabs the reader's interest. Or perhaps program evaluation and assessment are the reader's strengths. Apply intelligence and energy to that task. NIH staffers Lura Abbott and Christine Grady conducted a systematic review of the empirical literature evaluating IRBs, published in 2011. Grady, a bioethicist at NIH, and Abbott, a retired IRB Coordinator for NIH, conclude that the review protocols and regulations have not kept pace with research innovations. They analyze the findings of forty-three empirical studies and discover a lack of standardization. U.S. IRBs differ in their application of the federal regulations, in the time they take to review studies, and in the decisions made. None of the studies included an evaluation of IRB effectiveness, but they acknowledge this is needed. They recommend further research to understand how IRBs accomplish their objectives, what issues they find important, what defines and determines IRB review quality, and how effective IRBs are at protecting human research participants.

Friesen, Redman, and Caplan, ethicists at New York University (NYU), writing eight years after Abbott and Grady, address the increased complexity and commercialization of the IRB system. In the early days, IRBs tended to have one principal investigator, at one site, and were government-sponsored, clinical research. Currently, research is often privately funded, multisite, and multinational. The volume of trials has increased, and the research has become "more and more commercialized."[127] Friesen and her colleagues make some suggestions for reform. They advise there is the need for an appeals process, training needs to be revamped, and efficacy measures reconceived in relation to all regulatory bodies that are built on the IRB model. They suggest that the extent of the problem be realized, definitions harmonized, authority of reporting relationships clarified, and that this systemic overview be thoroughly documented.[128]

Human subjects are vulnerable to exploitation by the ambitious investigator, and the IRB model was devised to protect them. But sometimes, well-meaning protectors—the federal civil servants and the IRB—become entrenched in their roles and become the villains (in the eyes of some). Learned societies and DC lobbyists come to the rescue. But new heroes are needed, those with altruism and the ability to gather, analyze, and synthesize information in order to reform the IRB system. The history of the IRB is a genuinely engrossing narrative worthy of the information professional's time as it is truly pertinent to their work. Set against the scrim of scholarly communications, it involves the political economy of knowledge creation, the ethics of scientific experimentation, and perceived threats to the principles of academic freedom. Sounds like a job for librarians, and particularly academic librarians, wouldn't you agree?

CONCLUSION

IRBs are important players in the chain of scholarly communication. The IRB has a critical place in the creation of knowledge, falling somewhere between hypothesis formation and actual data gathering. IRBs are ethical gatekeepers, standard bearers for institutions that are doing human subject research. In their support for the information needs of their faculty and students, librarians provide services to enhance scholarly communication and research. They might find themselves in a supportive role, or in the crossfire, or on a crusade to make the IRB process better. They may simply keep the process rolling, enabling well-meaning agency bureaucrats. Or they may secretly side with the resistance, balancing a nuanced professional position and a different personal stance.

And now we return again to our multi-faceted/dimensional timeline with these threads: economic/political and moral/ethical. How is research (and the oversight thereof) funded and regulated? In learning the genesis story of the IRB, one begins to understand the complex political nature of the process. Politicians, civil servants, and lobbyists leverage and spend social capital while pursuing pragmatically idealistic legislation, policies, and regulations. To that mix, add the muckraking influence of investigative journalism and responsiveness to the resultant public outcry. The situation is further complicated by the tension between academia and federal government agencies. Academia's college- and departmental-oriented ethos of decentralized governance coexists and is often dependent upon, but clashes with, the centralized federal system of agencies disbursing government research grants.

The history of the IRB is a tale replete with villainous Nazis and genocidal yes-men, scientists blinded by intellectual greed, heroes in white coats, and assiduous bureaucrats. To contemplate the various narratives of IRB history is to comprehend the cultural discourse, to integrate the morality tale into your subconscious, and to be open to learning from the real-life fable. It serves to inoculate us against complacence, numbness, cynicism, compassion fatigue, and a closed heart and mind. As the activists' saying goes, "If you're not outraged, you're not paying attention." To look at the situation that faced the national commission—tasked with writing a code and a process that included protection of the most vulnerable of humanity—is to see our current challenges with new eyes that view obstacles as scalable and surmountable, and that view mistakes as reparable. It encourages and reawakens our natural impulse to make the world a better place. If we allow it to soak into our psyche, it takes our understanding to a new level. It has the potential to make us better at our jobs—to understand systemic cultural and historical causes of the current situation. And on a personal level it can cause us to examine and come to peace with our personal response, whether it be moral outrage, compassion fatigue, cynical apathy, or constructive action.

On a professional level, librarians are fortunate to be in a profession that presents opportunities to educate. Thus, we can seize every opportunity to employ Elie Wiesel's advice to educators in our work. Wiesel believes the legacy of Nuremberg is that it calls for a reexamination of not only medicine and human experimentation, but also of learning, education, and culture:

> The respect for human rights in human experimentation demands that we see persons as unique, as ends in themselves. . . . [W]hen we teach, we must not see any person as an abstraction. Instead, we must see in every person a universe with its own secrets, with its own treasures, with its own sources of anguish, and with some measure of triumph.[129]

Libraries are well-positioned to lead and participate in the reexamination of learning, education, and culture Wiesel encourages. Every interaction with a researcher (e.g., creation of an access point, finding aid, or online guide, even the simple recommendation of a book) contains this positive potential to teach. Every review of a research project is also an opportunity to teach and learn. Librarians have the opportunity to meet fellow human beings empathetically. The experience is paradoxical. It is macrocosmic in that the world is pluralistically diverse and microcosmic in that the diversity manifests in the unique individuality of each person. If we are fortunate enough to serve on an IRB or we are peripherally involved, or if we are employed at an institution that has IRBs, we can advance the process, theory, and outcome by learning all we can about IRB history and taking strategic, positive actions.

NOTES

1. The National Research Act required that IRBs at institutions receiving Public Health Service (PHS) grants review human subjects research projects, but did not specify whether IRBs must review the projects not funded by the PHS. This ambiguity resulted in an effective "foot in the door" strategy—most of the assurances (90 percent) filed with the Department of Health, Education, and Welfare (DHEW) promised to review all human subjects research, regardless of funding source (Schrag 42–43, 2010). Zachary M. Schrag, *Ethical Imperialism: Institutional Review Boards*

and the Social Sciences, 1965–2009, Baltimore, MD: Johns Hopkins University Press, 2010.

2. I once attended a workshop about dual degree librarianship (MLS and JD) at a copyright conference for librarians. The discussion veered toward pros and cons of going to law school. One of the pros was an increase in salary. The moderator asked the audience for a show of hands—was their motivation to become librarians financial? To state it plainly, had they become librarians for the monetary rewards? The audience laughed knowingly and not a person raised their hand.

3. Ezekiel Emanuel et al., editors, *Oxford Textbook of Clinical Research Ethics*, New York: Oxford University Press, 2008.

4. Historian Susan M. Reverby notes that on a rhetorical level, Tuskegee became "America's Nuremberg" and became the number one metaphor for evil in research. But she cautions that this causes the details of the study to enter a "historical fog" and meaningful questions about racism, healthcare, and research are tamped: "The linkage between Nazis and Tuskegee which took place especially when it became part of the begat stories told in an opening lecture in a professional ethics course or in a listing of abuses in an institutional review board training document can obscure rather than highlight what has happened" (Reverby 2009, 192–93). Susan M. Reverby, *Examining Tuskegee: The Infamous Syphilis Study and Its Legacy*, Chapel Hill: University of North Carolina Press, 2009.

5. Reverby deems Tuskegee a "flashbulb" memory, defined by cognitive scientists as "a highly significant event (as opposed to a long series of events) that gets remembered because it is constantly reinforced in discussions and in the media" (Reverby 2009, 193).

6. Anthropologist Elvi Whittaker reveals some of the historical discourses and identifies emerging stories as well in her exegesis of competing narratives with regard to research ethics in academia. She refers to the "discourses lurking in the cultural background such as the freedom of information and 'the people's right to know,'" themes with which librarians are well acquainted (Whittaker 2005, 514). Elvi Whittaker, "Adjudicating Entitlements: The Emerging Discourses of Research Ethics Boards," *Health: An Interdisciplinary Journal for the Social Study of Health, Illness, and Medicine* 9, no. 4 (2005): 513–35, DOI: 10.1177/1363459305056416.

7. Charles R. MacKay, "The Evolution of the Institutional Review Board: A Brief Overview of Its History," *Clinical Research and Regulatory Affairs* 12, no. 2 (1995): 65–94. DOI: 10.3109/10601339509079579.

8. Waldemar Kaempffert, "Science in Review: Genocide as Practiced by Nazis: Experiments With 'Dumb Cane,' Prisoners for Research," *New York Times* (November 17, 1946): E9.

9. Evelyne Shuster, "American Doctors at the Nuremberg Medical Trial," *American Journal of Public Health* 108, no.1 (January 2018): 47. Published online 2018 January. DOI: 10.2105/AJPH.2017.304104.

10. American Association for the Advancement of Science (AAAS), "Intersections of Science, Ethics and Human Rights: The Question of Human Subjects Protection: Report of the Science Ethics and Human Rights Working Group," AAAS Science and

Human Rights Coalition (February 2012): 3, https://www.aaas.org/sites/default/files/
s3fs-public/reports/ScienceEthicsHumanRights.pdf.

11. Arianne M. LaChapelle-Henry, Priyanka D. Jethwani, and Michael A. Grodin,
"The Complicated Legacy of the Nuremberg Code in the United States," in *Medi-
cal Ethics in the 70 Years after the Nuremberg Code, 1947 to Present*, edited by H.
Czech, C. Druml, and P. Weindling. Springer (2018), *Wien Klin Wochenschr: The
Central European Journal of Medicine* (2018) 130: S177, https://doi.org/10.1007/
s00508-018-1343-y.

12. Shuster, "American Doctors," 47.

13. Ibid., 51.

14. Ibid., 50.

15. LaChapelle-Henry et al., "The Complicated Legacy," S177.

16. Ibid., S180.

17. Ibid.

18. Ibid.

19. Erika Blacksher and Jonathan D. Moreno, "A History of Informed Consent in
Clinical Research," in *Oxford Textbook of Clinical Research Ethics*, edited by Ezekiel
Emanuel et al., New York: Oxford University Press, 2008. [This source includes an
extensive three-page timeline.]

20. American Medical Association (AMA), "Supplementary Report of the Judicial
Council," *Journal of the American Medical Association* (1946) 132: 1090.

21. Blacksher and Moreno, "A History of Informed Consent," 597.

22. MacKay, "The Evolution of the Institutional Review Board," 67–68.

23. Ibid., 67–76.

24. Ibid., 67.

25. Laura Stark, *Behind Closed Doors: IRBs and the Making of Ethical Research,*
Chicago: University of Chicago Press, 2012: 75–76.

26. MacKay, "The Evolution of the Institutional Review Board," 68.

27. U.S. Department of Health, Education, and Welfare (USDHEW), National Institutes of Health (NIH), 1953, Group Consideration of Research Procedures Deviating
from Accepted Medical Practice or Involving Unusual Hazard. [Hyperlinked at 1953
Milestone at https://history.nih.gov/display/history/Human+Subjects+Timeline.]

28. National Research Council (NRC) of the National Academies, Committee on
the Use of Third Party Toxicity Research with Human Research Participants, Science, Technology, Policy and Global Affairs Division, "The Regulatory Framework
for Protecting Humans in Research," 47, in *Intentional Human Dosing Studies for
EPA Regulatory Purposes: Scientific and Ethical Issues*. Washington, DC: National
Academies Press, 2004, https://www.ncbi.nlm.nih.gov/books/NBK215883/.

29. Stark, *Behind Closed Doors*, 76–77.

30. Curran as cited in Robert J. Levine, "Institutional Review Boards," *Bioethics*,
fourth ed., edited by Bruce Jennings, Farmington Hills, MI: Macmillan Reference,
2014, Gale eBooks.

31. National Research Council (NRC) of the National Academies, Committee on
the Use of Third-Party Toxicity Research with Human Research Participants, Sci-

ence, Technology, Policy and Global Affairs Division, "The Regulatory Framework," 46–47.

32. William J. Curran, "Governmental Regulation of the Use of Human Subjects in Medical Research: The Approaches of Two Agencies," *Daedalus* 98, no. 2 (Spring 1969): 545–46, http://www.jstor.com/stable/20023891.

33. Ibid., 546.

34. MacKay, "The Evolution of the Institutional Review Board," 70.

35. National Research Council (NRC) of the National Academies, Committee on the Use of Third-Party Toxicity Research with Human Research Participants, Science, Technology, Policy and Global Affairs Division, "The Regulatory Framework," 48.

36. Blacksher and Moreno, "A History of Informed Consent," 600.

37. Lawrence O. Gostin, "In Memorium: In Memory of William J. Curran, Francis Glessner Lee Professor of Legal Medicine, Harvard University," *Health & Human Rights* 2, no. 2 (1997): 5, accessed at https://cdn2.sph.harvard.edu/wp-content/uploads/sites/13/2014/03/3-Gostin.pdf.

38. Curran, "Governmental Regulation of the Use of Human Subjects," 542. MacKay's administrative history also remarks upon the influence of economics on this era. In the period 1950–1968, national expenditures for biomedical research increased more than fifteen-fold to $2.5 billion, with nearly 60 percent of that figure routed to NIH for research funding (Mackay 1995, 69).

39. Curran, "Governmental Regulation of the Use of Human Subjects," 588–89.

40. Ibid., 589.

41. Stark, *Behind Closed Doors*, 162–63.

42. Most of the encyclopedia entries that provide overviews of IRB history ignore or gloss over the economic context and the overarching core mission of academia to create knowledge and the competitiveness that engenders.

43. Schrag, *Ethical Imperialism,* 24.

44. Hugh Davis Graham and Nancy Diamond, *The Rise of American Research Universities: Elites and Challengers in the Postwar Era*, Baltimore, MD: Johns Hopkins University Press, 1997: 11.

45. Vannevar Bush and the Office of Scientific Research and Development (OSRD), *Science: The Endless Frontier* (1945), https://www.nsf.gov/about/history/EndlessFrontier_w.pdf.

46. See chapter 2 of *Science: The Endless Frontier*, titled "The War against Disease," at https://www.nsf.gov/od/lpa/nsf50/vbush1945.htm#ch2.

47. Margaret Foster Riley, "Federal Funding and the Institutional Evolution of Federal Regulation of Biomedical Research," *Harvard Law and Policy Review* 5 (2011): 267.

48. Graham and Diamond, *The Rise of American Research Universities*, 33.

49. Ibid., 34.

50. Ibid., 4.

51. MacKay, "The Evolution of the Institutional Review Board," 71.

52. Annette Flanagin, "Who Wrote the Declaration of Helsinki?" *Journal of the American Medical Association* 277, no. 11 (March 19, 1997): 926. DOI:10.1001/jama.1997.03540350076039.

53. Evelyne Shuster, "Fifty Years Later: The Significance of the Nuremberg Code," *New England Journal of Medicine* 337 (November 13,1997): 1440, DOI: 10.1056/NEJM199711133372006.

54. John A. Osmundsen, "Many Scientific Experts Condemn Ethics of Cancer Injection: Medical Research Circles Are Buzzing over the Disclosure Last Week of Experiments in Which Persons Were Injected with Living Cancer Cells without Their Knowledge," *New York Times* (January 26, 1964): 70.

55. David J. Rothman, *Strangers at the Bedside: A History of How Law and Ethics Transformed Medical Decision Making*, New York: Basic Books, 1991: 87.

56. Ibid., 73.

57. Ibid., 87.

58. Charles R. McCarthy, "The Origins and Policies That Govern Institutional Review Boards," 542, in *Oxford Textbook of Clinical Research Ethics*, edited by Ezekiel Emanuel et al., New York: Oxford University Press, 2008. [Includes a timeline of the development of federal protections for human subjects of research.]

59. U.S. Department of Health, Education, and Welfare (DHEW), PHS, Surgeon General's Directives on Human Experimentation, July 1, 1966 at 1966 policy statement of the U.S. Surgeon General policy statement ["Clinical research and investigation involving human beings," Surgeon General, Public Health Service to the Heads of the Institutions Conducting Research with Public Health Service Grants], https://history.nih.gov/display/history/Human+Subjects+Timeline.

60. David S. Jones and Christine Grady Lederer, "Ethics and Clinical Research: The 50th Anniversary of Beecher's Bombshell," *New England Journal of Medicine* 374, no. 24 (June 16, 2016): 2396.

61. Vincent J. Kopp, "Henry Knowles Beecher and the Development of Informed Consent in Anesthesia Research," *Anesthesiology* 90, no. 6 (June 1999): 1764.

62. Jones and Lederer, "Ethics and Clinical Research," 2397.

63. Rothman cited by Jones and Lederer, "Ethics and Clinical Research," 2396.

64. U.S. Department of Health and Human Services (DHHS), National Institutes of Health (NIH), Office of NIH History and Stetten Museum, *Timeline of Laws Related to the Protection of Human Subjects*, compiled by Joel Sparks, Bethesda, MD, June 2002, https://history.nih.gov/display/history/Human+Subjects+Timeline.

65. Robert A. Greenwald, Mary Kay Ryan, and James E. Mulvihill, editors, *Human Subjects Research: A Handbook for Institutional Review Boards*, New York: Plenum Press, 1982: 6.

66. David J. Rothman, *The Willowbrook Wars*, New York: Harper & Row, 1984.

67. National Research Council (NRC) of the National Academies, "Regulatory History," 63, in *Protecting Participants and Facilitating Social and Behavioral Sciences Research*, Washington, DC: The National Academies Press, 2003, available from https://doi.org/10.17226/10638 and https://www.nap.edu/download/10638#.

68. McCarthy, "The Origins and Policies That Govern Institutional Review Boards," 547.

69. Levine, "Institutional Review Boards," 1736.

70. U.S. Congress, National Research Act, Title II, Public Law 93-348, 1974, accessed at https://history.nih.gov/download/attachments/1016866/PL93-348.pdf.

71. Sarah L. Babb, *Regulating Human Research: IRBs from Peer Review to Compliance Bureaucracy,* Redwood City, CA: Stanford University Press, 2020: 19.

72. Ibid., 18–19.

73. Ibid., 20.

74. LaChapelle-Henry et al., "The Complicated Legacy"; Shuster, "American Doctors at the Nuremberg Medical Trial."

75. U.S. Congress, National Research Act, Title II, Public Law 93-348.

76. U.S. Department of Health and Human Services (DHHS), Office for Human Research Protections (OHRP) and Duane Alexander, Interview transcript, July 9, 2004. Interviewer: Patricia C. El-Hinnawy, https://bioethics.georgetown.edu/word press/wp-content/uploads/cnn4/InterviewAlexander.pdf.

77. Ibid.

78. McCarthy, "The Origins and Policies that Govern Institutional Review Boards," 548.

79. U.S. Department of Health and Human Services (DHHS), Office for Human Research Protections (OHRP), and Miriam Kelty, August 26, 2004; Kelty, Miriam Interview Transcript, https://bioethics.georgetown.edu/wordpress/wp-content/uploads/cnn4/InterviewKelty.pdf.

80. McCarthy, "The Origins and Policies that Govern Institutional Review Boards," 548.

81. Brian Callahan, *What Price Better Health? Hazards of the Research Imperative*, Berkeley: University of California Press, 2003.

82. U.S. Department of Health and Human Services (DHHS), Office for Human Research Protections (OHRP), and Charles R. McCarthy, Belmont Oral History Project / OHRP / Oral History of the Belmont Report and the National Commission for the Protection of Human Subjects of Biomedical and Behavioral Research / Interview with Charles R. McCarthy, PhD, NIH Liaison to the Commission, NIH, Bethesda, July 22 2004. Interviewer Patricia El Hinnawy, [Interview Transcript] https://bioeth ics.georgetown.edu/wordpress/wp-content/uploads/cnn4/InterviewMcCarthy.pdf.

83. Ibid.

84. Ibid.

85. Ibid.

86. Kennedy Institute of Ethics, Biomedical Research Library, "Library Materials," accessed May 27, 2021, at https://bioethics.georgetown.edu/library-materials/.

87. DHHS OHRP and Kelty, "Kelty, Miriam Interview Transcript."

88. Ibid.

89. Bradford H. Gray, Robert A. Cooke, and Arnold S. Tannenbaum, "Research Involving Human Subjects: The Performance of Institutional Review Boards Is Assessed in This Empirical Study," *Science* 201, no. 4361 (1978, September 22): 1101, http://www.jstor.org/stable/1746310.

90. MacKay, "The Evolution of the Institutional Review Board," 88.

91. McCarthy, "The Origins and Policies that Govern Institutional Review Boards," 549.

92. Babb, *Regulating Human Research*, 19–23.

93. Schrag, *Ethical Imperialism*, 190.

94. Ibid., 115.

95. Ibid., 54–77.

96. The tone of the discourse of the histories upon which I relied to write this chapter vary depending upon many factors, one of which seemed to be the professional affiliation of the author. The histories of the academicians tended to be highly critical and relied upon copious evidence (e.g., sociologists Babb and Stark, historian Schrag). The histories of the agency insiders (e.g., Mackay a Project Clearance Officer of the Office of Extramural Research at NIH; and McCarthy, former director of the Office for Prevention of Research Risks at NIH, and former liaison to the Commission) were less critical of bureaucracy and presented a more heroic narrative. In the civil servants' view, the regulatory mission was assumed as a given, and as a noble one that was not questioned. Ethicist Callahan and Harvard public intellectual Curran had balanced views, with Columbia University historian Rothman being the best storyteller. Bioethicist Erika Blacksher (University of Kansas) did not follow this bifurcation, as she was both critical and academic in tone but heroic in pursuing her aims of social justice, unafraid to be a gadfly toward psychologists with this cross-disciplinary jab in her timeline: "1952: American Psychological Association adopts ethics code that includes a strict consent requirement, which remained in tension with profession's routine use of deceit as a methodology" (Blacksher 2008, 593). To get the fullest impression of historical events, I read a wide swath of viewpoints, always considering the source. I leaned heavily on MacKay's tripartite view of IRB history to structure my own version of the history. His 1995 article provides the views of an agency insider and a pro-bureaucratic, carefully considered analysis of the dynamics of power. Librarians can appreciate his paean to documentation: "[H]istory and law teach us that the power of the document is considerable . . . records . . . promote autonomy and empowerment . . . thorough documentation of effective performance contributes not only to the perception of empowerment but to its reality as well" (Mackay 1997, 87–88). I considered adding Schrag's contentious "regimes" and Babb's "hyper-compliance" to the timeline to fairly represent resistance to what they perceive as unfair regulatory practices. But instead, I incorporated them into the narrative. And in the midst of this strife, I was glad for the respite of the Belmont Oral History Project, and I particularly enjoyed McCarthy's Belmont oral history interview. His enthusiasm for libraries, librarians, philosophy, ideas, and the collection and review of literature was uplifting and gratifying.

97. Schrag, *Ethical Imperialism*, 188–91.

98. Ibid., 187–92.

99. Zachary M. Schrag, "Institutional Review Blog," last updated July 2018, http://www.institutionalreviewblog.com/.

100. McCarthy, "The Origins and Policies that Govern Institutional Review Boards," 549.

101. Richard Speiglman and Patricia Spear, "The Role of Institutional Review Boards: Ethics: Now You See Them Now You Don't," in *Handbook of Social Research Ethics*, edited by Donna M. Mertens and Pauline E. Ginsberg, Thousand Oaks, CA: Sage, 2013: 4, https://dx.doi.org/10.4135/9781483348971.

102. National Research Council, "The Regulatory Framework for Protecting Humans in Research," 2004.

103. Ibid.

104. Ibid.

105. Eileen Welsome, *The Plutonium Files: America's Secret Medical Experiments in the Cold War*, New York: Dial Press, 1999: 482.

106. Ibid., 484.

107. Ibid., 487.

108. Ibid.

109. Ibid., 449.

110. Ibid., 460.

111. Ibid., 10.

112. Ibid., 492.

113. Ibid., 4.

114. American Library Association, "Professional Ethics," 2021a, http://www.ala.org/tools/ethics.

115. American Library Association, "Issues and Advocacy," 2021b, http://www.ala.org/advocacy.

116. Schrag, *Ethical Imperialism*, 130–34.

117. Ibid., 135.

118. Babb, *Regulating Human Research*, 26.

119. Ibid., 84.

120. Ibid., 31–62.

121. U.S. Department of Health and Human Services (DHHS), National Institutes of Health (NIH), National Institute of Environmental Health Sciences (NIEHS) and David B. Resnick, *Research Ethics Timeline*, Durham, NC, last reviewed August 25, 2020, https://www.niehs.nih.gov/research/resources/bioethics/timeline/index.cfm.

122. Carl Straumsheim, "No More 'Beall's List,'" *Inside Higher Education* (2017 January 18), https://www.insidehighered.com/news/2017/01/18/librarians-list-predatory-journals-reportedly-removed-due-threats-and-politics.

123. DHHS HIS NIEHS and Resnick, *Research Ethics Timeline.*

124. DHHS OHRP and Kelty, "Kelty, Miriam Interview Transcript."

125. Myra Parker, Cynthia Pearson, Caitlin Donald, and Celia B. Fisher, "Beyond the Belmont Principles: A Community–Based Approach to Developing an Indigenous Ethics Model and Curriculum for Training Health Researchers Working with American Indian and Alaska Native Communities," *American Journal of Community Psychology* 64, no. 1/2 (2019): 9, doi:10.1002/ajcp.12360.

126. Ibid.

127. Phoebe Friesen, Barbara Redman, and Arthur Kaplan, "Of Straw, Camels, Research Regulation, and IRBs," *Therapeutic Innovation and Regulatory Science* 53, no. 4: 526, DOI: 10.1177/2168479018783740.

128. Ibid., 531–32.

129. George J. Annas and Michael A. Grodin, eds., *The Nazi Doctors and the Nuremberg Code: Human Rights in Human Experimentation*, New York: Oxford University Press, 1992: ix.

BIBLIOGRAPHY

Abbott, Lura, and Christine Grady. "A Systematic Review of the Empirical Literature Evaluating IRBs: What We Know and What We Still Need to Learn." *Journal of Empirical Research of Human Research Ethics* 6, no. 1 (March 2011): 3–19.

American Association for the Advancement of Science. "Intersections of Science, Ethics and Human Rights: The Question of Human Subjects Protection: Report of the Science Ethics and Human Rights Working Group." AAAS Science and Human Rights Coalition. (February 2012). https://www.aaas.org/sites/default/files/ s3fs-public/reports/ScienceEthicsHumanRights.pdf.

American Association of University Professors (AAUP). Reports and Publications: "Regulation of Research on Human Subjects: Academic Freedom and the Institutional Review Board." (March 2013). https://www.aaup.org/file/IRB-Final-Report .pdf.

American Educational Research Association. "Policy and Advocacy: Issues and Initiatives: The Common Rule for the Protection of Human Subjects of Research," accessed at https://www.aera.net/Research-Policy-Advocacy/Issues-and-Initiatives/ The-Common-Rule-for-the-Protection-of-Human-Subjects-in-Research.

American Library Association. "Professional Ethics." 2021a. http://www.ala.org/ tools/ethics.

American Library Association. "Issues & Advocacy." 2021b. http://www.ala.org/ advocacy.

American Medical Association. "Supplementary Report of the Judicial Council." *Journal of the American Medical Association* 132 (1946):1090.

Annas, George J., and Michael A. Grodin, eds. *The Nazi Doctors and the Nuremberg Code: Human Rights in Human Experimentation*. New York: Oxford University Press, 1992.

Annas, George J., and Michael A. Grodin. "Reflections on the 70th Anniversary of the Nuremberg Doctors' Trial." *American Journal of Public Health* 108, no. 1 (January 2018): 10–12. DOI: 10.2105/AJPH.2017.304203.

Babb, Sarah L. *Regulating Human Research: IRBs from Peer Review to Compliance Bureaucracy*. Redwood City, CA: Stanford University Press, 2020.

Beecher, Henry K. "Experimentation in Man." *Journal of the American Medical Association* 169, no. 5 (1959): 461–78.

Beecher, Henry K. "Special Article: Ethics and Clinical Research." *New England Journal of Medicine* 274, no. 24 (1966): 1354–60. https://www.hhs.gov/ohrp/ regulations-and-policy/archived-materials/index.html.

Blacksher, Erika, and Jonathan D. Moreno. "A History of Informed Consent in Clinical Research." In *Oxford Textbook of Clinical Research Ethics*, 591–605, edited by Ezekiel Emanuel et al. New York: Oxford University Press, 2008.

Briggle, Adam. "Institutional Review Boards." In *Encyclopedia of Science, Technology and Ethics*. Detroit: Macmillan Reference, 2005.

Bush, Vannevar. Office of Scientific Research and Development. *Science: The Endless Frontier* (1945). https://www.nsf.gov/about/history/EndlessFrontier_w.pdf.

Callahan, Brian. *What Price Better Health? Hazards of the Research Imperative.* Berkeley: University of California Press, 2003.

Curran, William J. "Governmental Regulation of the Use of Human Subjects in Medical Research: The Approaches of Two Agencies." *Daedalus* 98, no. 2 (Spring 1969): 542–94. http://www.jstor.com/stable/20023891.

El-Hai, Jack. "H. K. Beecher: Brief Life of a Late Blooming Ethicist 1904–1976." *Harvard Magazine* (March-April 2017). Accessed at https://harvardmagazine.com/2017/03/henry-knowles-beecher.

Emanuel, Ezekiel, et al., editors. *Oxford Textbook of Clinical Research Ethics.* New York: Oxford University Press, 2008.

Flanagin, Annette. "Who Wrote the Declaration of Helsinki?" *Journal of the American Medical Association* 277, no. 11. (1997): 926. DOI:10.1001/jama.1997.03540350076039.

Friesen, Phoebe, Barbara Redman, and Arthur Kaplan. "Of Straw, Camels, Research Regulation, and IRBs." *Therapeutic Innovation and Regulatory Science* 53, no. 4 (2019): 526–34. DOI: 10.1177/2168479018783740.

Georgetown University, Kennedy Institute of Ethics, Bioethics Research Library. "History of the Bioethics Library," accessed at http://hdl.handle.net/10822/1043130.

Georgetown University, Kennedy Institute of Ethics, Bioethics Research Library. "Belmont Report Anniversary and History." [Virtual Exhibit]. Accessed at https://bioethics.georgetown.edu/library-materials/archives/belmont-report-anniversary-and-oral-history/.

Gostin, Lawrence O. "In Memorium: In Memory of William J. Curran, Francis Glessner Lee Professor of Legal Medicine, Harvard University." *Health & Human Rights* 2, no. 2 (1997): 5–8. Accessed at https://cdn2.sph.harvard.edu/wp-content/uploads/sites/13/2014/03/3-Gostin.pdf.

Grady, Christine. "Institutional Review Boards: Purposes and Challenges." *Chest* 148, no. 5 (November 2015): 1148–55. Published online 2015 Jun 4. DOI: 10.1378/chest.15-0706. [Includes table of U.S. regulatory requirements, timeline of regulations and guidance, and table of alternative models.]

Graham, Hugh Davis, and Nancy Diamond. *The Rise of American Research Universities: Elites and Challengers in the Postwar Era.* Baltimore, MD: Johns Hopkins University Press, 1997.

Gray, Bradford H., Robert A. Cooke, and Arnold S. Tannenbaum. "Research Involving Human Subjects: The Performance of Institutional Review Boards Is Assessed in This Empirical Study." *Science* 201, no. 4361 (1978): 1094–1101, http://www.jstor.org/stable/1746310.

Greenwald, Robert A., Mary Kay Ryan, and James E. Mulvihill, eds. *Human Subjects Research: A Handbook for Institutional Review Boards.* New York, Plenum Press, 1982.

Harris, Richard, *The Real Voice.* New York: Macmillan, 1964.

Jones, David S., and Christine Grady Lederer. "Ethics and Clinical Research: The 50th Anniversary of Beecher's Bombshell." *New England Journal of Medicine.* 374 no. 24 (June 16, 2016).

Kaempffert, Waldemar. "Science in Review: Genocide as Practiced by Nazis: Experiments With 'Dumb Cane,' Prisoners for Research." *New York Times* (17 November 1946): E9.

Katz, Jay, compiler. *Experimentation with Human Beings: The Authority of the Investigator, Subject, Professions, and State in the Human Experimentation Process*. New York: Russell Sage Foundation, 1972, pp. 9–65: The Jewish Chronic Disease Hospital Case, https://repository.library.georgetown.edu/handle/10822/1031369?show=full.

Kopp, Vincent J. "Henry Knowles Beecher and the Development of Informed Consent in Anesthesia Research." *Anesthesiology* 90, no. 6 (June 1999): 1756–65.

LaChapelle-Henry, Arianne M., Priyanka D. Jethwani, and Michael A. Grodin. "The Complicated Legacy of the Nuremberg Code in the United States." In *Medical Ethics in the 70 Years after the Nuremberg Code, 1947 to Present*. Edited by H. Czech, C. Druml, and P. Weindling. Springer (2018). *Wien Klin Wochenschr: The Central European Journal of Medicine* 130 (2018): S159–S253. https://doi.org/10.1007/s00508-018-1343-y.

Levine, Robert J. "Institutional Review Boards." In *Bioethics*, fourth ed. Edited by Bruce Jennings. Farmington Hills, MI: Macmillan Reference, 2014. Gale eBooks.

Library of Congress. "Military Legal Sources, Nuremberg Trial," 2021. Published June 4, 2014. https://www.loc.gov/rr/frd/Military_Law/Nuremberg_trials.html.

MacKay, Charles R. "The Evolution of the Institutional Review Board: A Brief Overview of Its History." *Clinical Research and Regulatory Affairs* 12, no. 2 (1995): 65–94. DOI: 10.3109/10601339509079579.

McCarthy, Charles R. "The Origins and Policies that Govern Institutional Review Boards." In *Oxford Textbook of Clinical Research Ethics* (pp. 541–51). Edited by Ezekiel Emanuel et al., New York: Oxford University Press, 2008. [Includes a timeline of the development of federal protections for human subjects of research.]

Mintz, Morton. *By Prescription Only*. Beacon Press, 1967.

National Research Council of the National Academies, Committee on the Use of Third-Party Toxicity Research with Human Research Participants. Science, Technology, Policy and Global Affairs Division. "The Regulatory Framework for Protecting Humans in Research" (pp. 46–65). In *Intentional Human Dosing Studies for EPA Regulatory Purposes: Scientific and Ethical Issues*. Washington, DC: National Academies Press, 2004. https://www.ncbi.nlm.nih.gov/books/NBK215883/.

National Research Council of the National Academies. "Regulatory History" (pp. 59–79). In *Protecting Participants and Facilitating Social and Behavioral Sciences Research*. Washington, DC: The National Academies Press, 2003. Available from https://doi.org/10.17226/10638. https://www.nap.edu/download/10638#.

National Research Council. Committee on Revisions to the Common Rule for the Protection of Human Subjects in Research in the Behavioral and Social Sciences. Board on Behavioral, Cognitive, and Sensory Sciences. Committee on National Statistics. Committee on Population; Division of Behavioral and Social Sciences and Education. "Proposed Revisions to the Common Rule for the Protection of Human Subjects in the Behavioral and Social Sciences." Washington, DC: National

Academies Press (US). (2014, March 31). 6, Improving the IRB Process. https://www.ncbi.nlm.nih.gov/books/NBK217974/.

Osmundsen, John A. "Many Scientific Experts Condemn Ethics of Cancer Injection." *New York Times* (26 January 1964): 70.

Parker, Myra, Cynthia Pearson, Caitlin Donald, and Celia B. Fisher. "Beyond the Belmont Principles: A Community–Based Approach to Developing an Indigenous Ethics Model and Curriculum for Training Health Researchers Working with American Indian and Alaska Native Communities." *American Journal of Community Psychology* 64, no. 1/2 (2019): 9–20. doi:10.1002/ajcp.12360.

Reverby, Susan M. *Examining Tuskegee: The Infamous Syphilis Study and Its Legacy.* Chapel Hill: University of North Carolina Press, 2009.

Riley, Margaret Foster. "Federal Funding and the Institutional Evolution of Federal Regulation of Biomedical Research." *Harvard Law and Policy Review* 5 (2011): 265–87.

Rothman, David J. *Strangers at the Bedside: A History of How Law and Ethics Transformed Medical Decision Making.* New York: Basic Books, 1991.

Rothman, David J. *The Willowbrook Wars.* New York: Harper & Row, 1984.

Schrag, Zachary M. *Ethical Imperialism: Institutional Review Boards and the Social Sciences, 1965–2009.* Baltimore, MD: Johns Hopkins University Press, 2010.

Schrag, Zachary M. "Institutional Review Blog." Last updated July 2018. http://www.institutionalreviewblog.com/.

Shuster, Evelyne. "American Doctors at the Nuremberg Medical Trial." *American Journal of Public Health* 108, no.1 (January 2018): 47–52. DOI: 10.2105/AJPH.2017.304104.

Shuster, Evelyne. "Fifty Years Later: The Significance of the Nuremberg Code." *New England Journal of Medicine* 337 (1997): 1436–40. DOI: 10.1056/NEJM199711133372006.

Speiglman, Richard, and Patricia Spear. "The Role of Institutional Review Boards: Ethics: Now You See Them Now You Don't." In *Handbook of Social Research Ethics* (pp. 121–34). Edited by Donna M. Mertens and Pauline E. Ginsberg. Thousand Oaks, CA: Sage, 2013. https://dx.doi.org/10.4135/9781483348971.

Stark, Laura. *Behind Closed Doors: IRBs and the Making of Ethical Research.* Chicago: University of Chicago Press, 2012.

Straumsheim, Carl. "No More 'Beall's List.'" *Inside Higher Education* (2017, January 18). https://www.insidehighered.com/news/2017/01/18/librarians-list-predatory-journals-reportedly-removed-due-threats-and-politics.

U.S. Congress, National Research Act, Title II, Public Law 93-348 (1974). Accessed at https://history.nih.gov/download/attachments/1016866/PL93-348.pdf.

U.S. Department of Energy (DOE). "DOE Openness: Human Radiation Experiment, Roadmap to the Project." https://ehss.energy.gov/ohre/roadmap/index.html.

U.S. Department of Health, Education, and Welfare (DHEW), and the National Institutes of Health (NIH). "Group Consideration of Research Procedures Deviating from Accepted Medical Practice or Involving Unusual Hazard." (1953). [Hyperlinked at 1953 Milestone at https://history.nih.gov/display/history/Human+Subjects+Timeline.]

U.S. Department of Health, Education, and Welfare, National Commission for the Protection of Human Subjects of Biomedical and Behavioral Research. "Report and Recommendations: Institutional Review Boards," DHEW Publication No. (OS) 78-0008." (September 1, 1978) Accessed at https://repository.library.georgetown .edu/bitstream/handle/10822/778625/ohrp_institutional_review_boards_1978 .pdf#page=1.

U.S. Department of Health, Education, and Welfare (DHEW). NIH policies on human subject protection, Federal Register, 39 105 part 2 (45 CFR 46), (May 30, 1974). https://tile.loc.gov/storage-services/service/ll/fedreg/fr039/fr039105/fr039105.pdf.

U.S. Department of Health, Education, and Welfare (DHEW), Public Health Service (PHS), "Surgeon General's Directives on Human Experimentation, July 1 1966 at 1966 policy statement of the U.S. Surgeon General policy statement." ["Clinical research and investigation involving human beings," Surgeon General, Public Health Service to the Heads of the Institutions Conducting Research with Public Health Service Grants.] https://history.nih.gov/display/history/ Human+Subjects+Timeline.

U.S. Department of Health and Human Services (DHHS), National Institutes of Health (NIH), Office of NIH History and Stetten Museum, *Timeline of Laws Related to the Protection of Human Subjects* [compiled by Joel Sparks]. (June 2002.) https://history.nih.gov/display/history/Human+Subjects+Timeline.

U.S. Department of Health and Human Services (DHHS), National Institutes of Health (NIH), National Institute of Environmental Health Sciences (NIEHS). *Research Ethics Timeline* (by David B. Resnick). Last reviewed August 25, 2020. https://www.niehs.nih.gov/research/resources/bioethics/timeline/index.cfm.

U.S. Department of Health and Human Services (DHHS), Office for Human Research Protections (OHRP), and Duane Alexander. Interview Transcript July 9, 2004 (Interviewer: Patricia C. El-Hinnawy). https://bioethics.georgetown.edu/wordpress/ wp-content/uploads/cnn4/InterviewAlexander.pdf.

U.S. Department of Health and Human Services (DHHS), Office for Human Research Protections (OHRP), and Miriam Kelty. Miriam Kelty Interview Transcript August 26, 2004. https://bioethics.georgetown.edu/wordpress/wp-content/uploads/cnn4/ InterviewKelty.pdf.

U.S. Department of Health and Human Services (DHHS), Office for Human Research Protections (OHRP), and Charles R. McCarthy. "Belmont Oral History Project / OHRP / Oral History of the Belmont Report and the National Commission for the Protection of Human Subjects of Biomedical and Behavioral Research" (Interview with Charles R. McCarthy, July 22, 2004, interviewer Patricia El. Hinnawy). Interview transcript. https://bioethics.georgetown.edu/wordpress/wp-content/uploads/ cnn4/InterviewMcCarthy.pdf.

U.S. Department of Health and Human Services (DHHS), Office for Human Research Protections (OHRP). History [timeline]. https://www.hhs.gov/ohrp/about-ohrp/his tory/index.html.

U.S. Department of Health and Human Services (DHHS), Office for Human Research Protections (OHRP). "The Belmont Report: Ethical Principles and Guidelines for the Protection of Human Subjects of Research; Related Historical Documents from the National Commission for the Protection of Human Subjects of Biomedical and Behavioral Research." https://www.hhs.gov/ohrp/regulations-and-policy/belmont-report/access-other-reports-by-the-national-commission/index.html.

Welsome, Eileen. *The Plutonium Files: America's Secret Medical Experiments in the Cold War*. New York: Dial Press, 1999.

Whittaker, Elvi. "Adjudicating Entitlements: The Emerging Discourses of Research Ethics Boards." *Health: An Interdisciplinary Journal for the Social Study of Health, Illness, and Medicine* 9, no. 4 (2005): 513–35. DOI: 10.1177/1363459305056416.

World Medical Association (WMA). "Declaration of Helsinki—Ethical Principles for Medical Research Regarding Human Subjects." Adopted by the 18th WMA General Assembly, Helsinki, Finland, June 1964. https://www.wma.net/wp-content/uploads/2018/07/DoH-Jun1964.pdf.

Chapter Three

Librarians and Institutional Review Boards

Roles, Responsibilities, and Searching Best Practices

Tracy C. Shields and Esther May Sarino

Institutional review boards (IRBs) are committees that are formally designated to review and monitor research involving human subjects.[1] In 1978, the National Commission for the Protection of Human Subjects of Biomedical and Behavioral Research released *The Belmont Report*, which summarizes the basic ethical principles of human subject research and offers a framework for IRBs to follow.[2] As discussed in chapter 1, the protection of human research subjects can be supported by librarians in various ways, from serving directly on IRBs to the services they provide to researchers and research participants. Librarians can be essential collaborators in the research process and their work can uphold the three core principles of *The Belmont Report*: Respect for Persons, Beneficence, and Justice.[3] For example, as nonscientist members of IRBs, librarians may stand as proxy representatives of research subjects, protecting their rights under the Respect for Persons principle. Similarly, they may support Beneficence and Justice with mediated searches so that the most appropriate literature to conduct research ethically is found and incorporated by researchers.

There is a paucity of published literature discussing roles, offering guidance, or suggesting best practices for librarians involved in IRBs. We use "librarian" as a catch-all term for anyone whose job position is based in a library or information center affiliated with a hospital or medical center, academic institution, corporation, or research setting. They may have a variety of job titles (e.g., reference librarian, research librarian, technical services coordinator, informationist, etc.) and be in a research, translational, bioinformatics, clinical, medical, or health sciences focused role.[4]

This chapter will touch on the various roles and responsibilities librarians in a medical or health sciences setting may have working with IRBs and researchers seeking IRB approval, along with some searching best practices based on the experiences of the authors and others involved in IRB concerns. Building on the roles outlined in chapter 1, this chapter will expand the discussion to other responsibilities that librarians can have with IRBs and emphasize how they may support research endeavors with comprehensive literature searching. Rather than a discussion of the IRB process from a *librarian as researcher* perspective—that is, how a librarian conducting research might seek IRB approval and navigate the regulatory submission process[5]—we discuss things from a *librarian as research advocate* and *librarian as consultant and/or member* viewpoints,[6] with an emphasis on searching best practices. It should be clear though that librarians may offer more to the research process than expert searching: librarians may be vital throughout the research lifecycle—planning and implementing research, working with grants, data management, preparing manuscripts for publishing when research is completed, and so on—and librarian involvement in IRBs can better inform them on how to meet researchers where they are in the process.

Librarians have the potential to play an important role in the IRB by developing a good connection with the department and becoming familiar with their processes, such as protocol submittals and renewals. Participating in IRBs or doing searches for incoming protocols are ways to work alongside the IRB administration.[7] The number of librarians directly involved in IRBs is unknown but assumed to be small. The emphasis is to integrate library services within the IRB process to establish a good working relationship between librarians, researchers, and regulatory bodies.

LIBRARIANS AND IRBS:
DIRECT AND INDIRECT INVOLVEMENT

It is not necessary for a librarian to be a voting member of the IRB to have a significant impact on IRBs and related research activities. The setting may dictate the approach and depth and involvement in IRB duties for librarians. For example, in academic environments, librarians with scientific backgrounds may serve as full scientific members, while librarians in a hospital setting may serve as nonscientist voting members.[8] Librarians may also serve on IRBs for other organizations outside of their work institution as nonaffiliated community members, as ex-officio members, or as consultants for scientific review committees in indirect support of IRBs.[9]

Case Example: The EVMS Experience

Eastern Virginia Medical School (EVMS) began their IRB librarian program in 2001 in response to the Johns Hopkins incident (see section below).[10]

Their approach was a collaboration with the campus IRB. An IRB librarian received and reviewed protocols a week ahead of the actual board meeting, presenting notes to the necessary personnel within the board and IRB administration reviewing the protocols in a timely manner.

Librarians at EVMS are still not officially included within the standard of protocols of the campus IRB and are not voting members, but they continue to be involved in sharing findings from searches and keeping constant communication with the IRB administration. Librarians maintain the relationship with the IRB by attending research efforts and training provided by the IRB department. The key approach is to become involved in multiple facets of the IRB. It is up to the librarian to be proactive in finding ways to partner with IRBs to support ongoing research. Continued engagement moves the librarians closer to becoming included in the regulatory process, widens the impact of the library as a whole, and elevates the library's profile beyond clinical, educational, and research support.

Other Roles

Outside of direct membership or involvement, librarians may offer other services to support IRBs and research endeavors, including comprehensive, mediated literature searching and training; clinical trial registration; discovery of funding sources and grant applications; health literacy instruction and consultation for informed consent; research data management; establishment and maintenance of institutional repositories; facilitating publication after completion in peer-reviewed journals; and lending information expertise at all points of the research life cycle.[11] The various roles demonstrate the added value of having a librarian on the IRB.[12]

The relationship between librarians and IRBs has benefits working both ways and may help inform and lead to the development of new library services and opportunities. For example, an IRB may have concerns about a researcher's plagiarism, a type of research misconduct, and require consultation with librarians on citation best practices. Librarians may expand education and training to that researcher on proper attribution, how to seek permissions, and other issues related to copyright and intellectual property. These efforts could be expanded to other researchers or become part of graduate medical education instruction. Librarians familiar with the inner workings of the IRB process may be key facilitators for new or early-career researchers. Acting as an "information hub," these librarians may connect trainees to mentors or suggest other professional connections gathered from their IRB experiences, help with writing or publishing their research, or be a "safe" source when they have questions about research but are hesitant to ask colleagues.

A major hurdle in librarians' involvement in IRBs is that IRB members and administrators may not see the need for librarians since they are not considered medical experts. However, librarians should be considered experts in finding information and searching the literature—two skill sets needed to prevent or reduce harm to human research subjects. Relying on discussions with colleagues to confirm that all relevant literature has been found is not enough. Librarians as third-party experts in the biomedical literature should not be overlooked. Nowhere was this clearer than in the Johns Hopkins University research study in which a healthy volunteer died because of information gaps and missing literature.

The Johns Hopkins Tragedy

In early 2001, Johns Hopkins University began "a study designed to provoke a mild asthma attack in order to help doctors discover the reflex that protects the lungs of healthy people against asthma attacks."[13] Subjects would inhale hexamethonium, a medication used in the 1950s and 1960s to treat high blood pressure, to induce the mild attack. As required by law, the research protocol was approved by an established IRB at Johns Hopkins and recruitment began.

Ellen Roche, a twenty-four-year-old technician from the Johns Hopkins Asthma and Allergy Center, volunteered as a participant. She was the third participant in the study. A day after she received the study drug, she developed a cough and became ill enough that she sought medical treatment and was put on a ventilator. She died about a month after entering the study.[14]

An investigation into Ms. Roche's death in relation to the research study soon followed.[15] In a report about the investigation, the university said the researcher who conducted the study and the IRB that approved it had failed to protect research subjects and take precautions to avoid harm.[16] Additionally, the Office for Human Research Protections found that the researcher and IRB "failed to obtain published literature about the known association between hexamethonium and lung toxicity" noting that "[s]uch data was readily available via routine MEDLINE and Internet database searches, as well as recent textbooks on pathology of the lung."[17]

As the Johns Hopkins case shows, an inadequate PubMed/MEDLINE search may miss important literature; a single resource, even if it is considered the "gold standard," may not be comprehensive enough and multiple sources should be used. IRB members may be overburdened with complex protocols or not have the depth of knowledge needed to understand the problems of a research study.[18] Additionally, researchers and IRB members may not have the information skills to assess the completeness or thoroughness of the literature review or biases against older literature—a factor that may have played a role in the Johns Hopkins case. Having an expert searcher, such as a librarian, involved with the IRB could address those biases and potential pitfalls.

THE IMPORTANCE OF THE LITERATURE SEARCH

While there were multiple factors that contributed to the Johns Hopkins tragedy, the report's highlight of an inadequate literature search was an opportunity for librarians to get involved in IRBs. As Tomlin noted, "Tempting as it is to say, 'Aha! They really do NEED us!' there's more to this issue than seeing our professional skill validated."[19] In the aftermath of the Johns Hopkins case, many librarians offered their services as information experts to IRBs and researchers, becoming partners in research protections.[20]

Unlike Institutional Animal Care and Use Committee (IACUC; the animal side of research regulatory bodies) regulations (see part II of this book), there are no set requirements for literature searches and reviews for IRBs, other than a generalized suggestion of a "comprehensive literature search" approach. Although *The Belmont Report* makes no specific reference to literature searches (librarian-mediated or not), a good, thorough literature search would certainly aid in the decision-making process of research and protect human subjects as much as possible.

Lack of consensus and different expectations for the literature review in the protocol submission process from regulatory bodies could lead to potential harm to human research subjects. IRBs may have different ideas of what a comprehensive literature search should be. Librarians as members or as outside consultants may offer IRBs guidance in literature search best practices and resource selection. These librarians should be comfortable with that type of collaboration and have the necessary skill sets and experiences with searching to demonstrate their expertise. Not all librarians will have those abilities—a cataloging librarian may be an excellent IRB nonscientist member but perhaps less ideal as an expert searcher than their reference librarian colleague.

Literature searches provide appropriate background information, flush out duplication of effort, identify gaps in literature for areas of potential research, and can help researchers hone into their research topic during protocol development. Discovery of adverse effects or potential harm should be the forefront of literature searches for human-based research studies. A thorough literature search—that is, a search that incorporates peer-reviewed literature without unnecessary date limitations across multiple resources, along with grey literature sources when appropriate—will find the necessary information in discovering negative effects a treatment might cause. The right literature can help assess the risks and benefits to the human subject. Comprehensive literature searches can also shed light on disparities in research, especially in under-represented groups or vulnerable populations. Librarians—especially those who do extensive searching and have a broad awareness of databases, specialized resources, and grey literature—may offer invaluable contributions and establish best practice to research endeavors.

Librarians as Search Experts

Librarians are familiar with users' information-seeking behaviors, including those of researchers. According to Haines and colleagues, "Traditional library services such as mediated literature searching and instruction" were rarely used by basic science researchers.[21] When researchers use the same methods and resources for each of their research projects without regard to nuances and specialized resources, it can lead to missing important information and literature.

Researchers may forego librarian-led training due to lack of time and experience with new resources. Obtaining information easily through electronic resources can discourage researchers from physically visiting the library or talking with a librarian. Lack of knowledge of the services the library has to offer and lack of acknowledgement of expertise is an ongoing issue for librarians. Having an expert searcher guide them through the process of searching will not only educate them on how to use resources, but will also help them to see the bigger picture of how librarians manipulate searches within the databases to retrieve necessary information. Librarians can perform mediated searches to generate better results for the researcher.

Librarian-Mediated Searches

Librarians can guide researchers, especially those new to the process, with their topic and help them refine their research question. As noted in part II, IACUC regulations include searching guidelines for duplication and alternatives, and protocols usually require some type of documentation of not only what resources were searched, but what terms or strategies were used and when they were searched. IRB regulations only speak to an assumption that a literature review is done and that it is comprehensive. As the Johns Hopkins case demonstrated, those assumptions are not always safe. Librarians may want to consider IACUC guidelines on duplicate and alternatives searches along with systematic review search documentation as conceptual templates for IRB-related comprehensive searches. Some IRBs may have their own requirements or guidance for literature searches, but these may be highly variable.

One problem librarians see in protocols, particularly those in which they have not provided assistance to the researchers, is a lack of in-depth searching. Literature significant to the research can be overlooked if an inadequate search is conducted. There is no current standard of protocol for a literature search to be included in the research submission. However, keeping record of the search process for the literature review may provide needed context on how and why the search was conducted in a specific way. This form of transparency will make it easy for those involved in the research and the IRB to see the structure of how references were found and selected for the topic. Evidence a search was conducted will show researchers did their due diligence in choosing their literature to support their research wisely.

When doing the search for the literature review, it might be good to do so in a systematic fashion, so researchers and IRB administrators can replicate the searches on their own. The librarian can educate IRB staff and reviewers on how to search for topics. The medical expert can work with the librarian in harvesting terms appropriate for their topic. Working together will ensure the literature is searched systematically.

To be clear, we do not mean to say that IRB searches should be equated with systematic review searches, although we do think IRB searches benefit from being done in a "systematic" way. While this systematic approach is not the same as systematic review searching, guidance and best practices surrounding systematic review searching can inform IRB searching. For example, systematic review search best practices, including use of multiple resources and clear documentation for reproducibility, may provide some structure and guidance for IRB-focused comprehensive searches. Librarians can document search strategies, the database (and interface or vendor version) used, the date of search, the number of results, and any filters or limits used for IRB protocol searching; the PRISMA-S checklist offers some structure in search strategy documentation that may inform IRB-related comprehensive searches.[22] This may seem excessive, but it benefits the researcher as well. Even if not required for IRB administrative paperwork or in a protocol, that information is helpful for shared duties in large research groups, updating literature for continuing reviews, or writing the manuscript and publishing when research is completed.

A habit of documentation is imperative: record everything properly and ensure search strategies are replicable for researchers and those reviewing the protocols. The contribution of detailed searches and search strategies are the useful data that enables the researcher to move forward on their topic and gives the IRB an overall look at the literature available. This will also aid future researchers attempting to study the same topic. Again, IACUC searching may offer some guidance to IRB-related searching by noting the different purposes of searches (e.g., duplication of research versus harms and adverse effects searches).

Where to Search

In many cases, the research question, treatment or intervention, and patient population will dictate the selection of resources to search. Librarians may have awareness of subject specific resources (free or paid subscription) that could greatly benefit researchers. Library subscriptions and changes in vendors are likely to impact searches as well. Experienced researchers may be competent searchers in MEDLINE, but perhaps not in the interface currently available to their institution; researchers may also not understand that one institution's offerings are not the same as another institution's subscriptions, and not account for changes in coverage, access, or availability. Nuances and differences in interfaces could greatly affect search results and the comprehensiveness of a literature search. A savvy librarian looking to expand their reach to researchers may consider giving the IRB a demonstration on how the same search in the same resource (e.g., MEDLINE) might give different results when done in different interfaces (e.g., PubMed versus Ovid's MEDLINE versus Ebsco's MEDLINE), possibly missing key literature, and emphasize that librarian involvement could mitigate these pitfalls.

A good rule of thumb is to search at least two to three different resources and add others as needed based on the topic, population, and so on. For example, a research study comparing two different corticosteroids in children with asthma exacerbation may need searches in PubMed/MEDLINE (as a starting point), Embase (for drug interactions), and grey literature sources such as Drugs@FDA (for warnings or approval documents). Other resources that may be informative include Web of Science (or Google Scholar) for cited reference searches, PsycINFO for social sciences, CINAHL for nursing topics, and the Defense Technical Information Center (DTIC) for military research.

Updating Searches

IRB continuing review offers researchers an opportunity to update the literature search and maintain awareness of ongoing trends in their research areas. Librarians are familiar with the ever-changing language of databases, and it is crucial the search strategies align with previous and current terms (which may or may not be the same terms accepted in the literature itself) in relation to the research topic. Controlled vocabulary may evolve (e.g., "Evaluation Studies"[MeSH] changed to "Evaluation Study"[MeSH]), new terms and concepts may arise, and resources may undergo redesigns (e.g., functional differences between legacy versus new PubMed). Librarians should keep abreast of these changes and shifts in information sources.

There needs to be emphasis on keeping searches updated for studies in relation to continuing reviews/ongoing research. Librarians can rerun the original search to find new published literature since the initial date the search was conducted. This is where documentation can be essential, not only for reproducibility but for continuity. Librarians should also check with the researcher to see if new concepts have been added to their scope of study to adjust the search strategy or create a possible new one for the change.

Updating relevant searches also catches new, related studies that may be of importance to the research. Many databases offer features to automate alerts for new literature through saved searches; third-party research tools may also be helpful to researchers. Librarians can educate researchers on options available to them through institutional subscriptions or other library holdings. These value-added methods are a possible way for librarians to market their services in new ways to different groups at their institution.

OTHER CONSIDERATIONS

When getting involved with an IRB, whether directly as a member or indirectly in supporting research efforts, librarians should consider the time and effort required. Professional duties and job responsibilities may dictate the type and/or level of involvement librarians may have with IRBs. Time commitments can be quite variable depending on an institution's research focus and the number of protocols. For IRB members, preparation for meetings includes reading protocols and any background information. Participation may prove to be a considerable obligation, based on the frequency with which the IRB meets and the length of those meetings, which may be in-person, virtual, or a mix.

Outside of those time commitments, training and continuing education are essential components for all IRB members. The Collaborative Institutional Training Initiative (CITI) Program offers training on research ethics and compliance with regulations on their website.[23] This training provides information on human subjects research and the history behind all the laws, issues, and regulations to protect human subjects. Becoming familiar with these topics will provide librarians interested in IRB roles better understanding on the importance of doing a thorough literature review and issues around conducting research with human subjects. Many institutions require ongoing certification from CITI for researchers, IRB members, and IRB consultants.

Librarians as IRB members will have continuing education as part of those duties, but librarians in other IRB-related roles will benefit from keeping up knowledge, skills, and abilities. Webinars, tutorials, and online courses from professional organizations may fill in gaps and expand skills for searching the literature. Unfortunately, there are few options specific to librarians on IRBs, so continuing education may mean doing a variety of things (e.g., staying current with database changes and search language, seeking guidance from fellow librarians on the latest tips/trends for searching, reading literature in support of searching skills) and seeing how the knowledge can be applied to the IRB librarian role.

In working on this chapter, we discovered gaps in professional opportunities for librarians who are involved in IRBs (directly or indirectly) to learn, connect, and share their experiences with other librarians in similar roles specific to the context of IRBs. While there are listservs and groups in professional organizations for librarians involved in IACUCs such as the Medical Library Association's Animal and Veterinary Information Specialist Caucus, we are not aware of any groups specifically focused on the needs of librarians as IRB members or affiliates.

CONCLUSION

Librarians can be involved in institutional regulatory boards like the IRB in various ways, from education to bolstering research. Librarians have a variety of skills and areas of knowledge to support research in human subjects. As IRB members, librarians can offer expertise in searching and health literacy that will benefit IRBs. Building relationships with the IRB helps expand connections with the research community and encourages investigators to include librarians as a part of the research process or team. We hope this chapter raises awareness of the potential collaborations between IRBs and librarians in different settings and leads to expanding their roles with IRBs.

NOTES

1. Richard Wagner, "Ethical Review of Research Involving Human Subjects: When and Why Is IRB Review Necessary?" *Muscle Nerve* 28, no. 1 (July 2003): 27–39. https://doi.org/10.1002/mus.10398.

2. National Commission for the Protection of Human Subjects of Biomedical and Behavioral Research, Department of Health, Education and Welfare (DHEW), "The Belmont Report." Washington, DC: United States Government Printing Office, September 30, 1978. https://www.hhs.gov/ohrp/regulations-and-policy/belmont-report/read-the-belmont-report/index.html.

3. Ibid.

4. Sarah Visintini, Mish Boutet, Alison Manley, and Melissa Helwig, "Research Support in Health Sciences Libraries: A Scoping Review," *Journal of the*

Canadian Health Libraries Association 39, no. 2 (July 24, 2018): 56–78. https://do
i.org/10.29173/jchla29366.

5. Robert V. Labaree, "Working Successfully with Your Institutional Review
Board: Practical Advice for Academic Librarians," *College & Research Libraries
News* 71, no. 4 (April 1, 2010): 190–93. https://crln.arcrl.org/index.php/crlnews/
article/view/8353/8494. Maura A. Smale, "Demystifying the IRB: Human Subjects
Research in Academic Libraries," *portal: Libraries and the Academy* 10, no. 3 (2010)
(3): 309–21. https://doi.org/10.1353/pla.0.0114.

6. Laureen Cantwell and Doris Van Kampen-Breit, "Librarians and the Institutional Review Board (IRB): Relationships Matter," *Collaborative Librarianship* 7,
no. 2 (2015): 66–78.

7. Labaree, "Working Successfully," 192.

8. Katherine Stemmer Frumento and Judith Keating, "The Role of the Hospital
Librarian on an Institutional Review Board," *Journal of Hospital Librarianship* 7, no.
4 (December 2007): 113–20. https://doi.org/10.1300/J186v07n04_08.

9. Cantwell and Van Kampen-Breit, "Librarians and the IRB," 71; Judith G. Robinson and Jessica Lipscomb Gehle, "Medical Research and the Institutional Review
Board: The Librarian's Role in Human Subject Testing," *Reference Services Review*
33, no. 1 (January 1, 2005): 20–24. https://doi.org/10.1108/00907320410519360.

10. Ibid.

11. Fern M. Cheek, "Research Support in an Academic Medical Center,"
Medical Reference Services Quarterly 29, no. 1 (January 2010): 37–46. https://doi
.org/10.1080/02763860903485068; Lisa Federer, "The Librarian as Research Informationist: A Case Study," *Journal of the Medical Library Association: JMLA* 101,
no. 4 (October 2013): 298–302. https://doi.org/10.3163/1536-5050.101.4.011; Paula
G. Raimondo, Ryan L. Harris, Michele Nance, and Everly D. Brown, "Health Literacy and Consent Forms: Librarians Support Research on Human Subjects," *Journal
of the Medical Library Association: JMLA* 102, no. 1 (January 2014): 5–8. https://doi
.org/10.3163/1536-5050.102.1.003; Brenda Fay, Jennifer Deal, and Vicki Budzisz,
"An Institutional Repository Experience at a Large Health Care System," *Medical
Reference Services Quarterly* 36, no. 3 (July 3, 2017): 280–91. https://doi.org/10.108
0/02763869.2017.1332264; Visintini et al., "Research Support," 63–66.

12. Sally Harvey, "Institutional Review Boards: Another Way for Hospital Librarians to Add Value to Their Organization," *Journal of Hospital Librarianship* 3, no. 2
(March 2003): 99–102. https://doi.org/10.1300/J186v03n02_09.

13. Julien Savulescu and Merle Spriggs, "The Hexamethonium Asthma Study and
the Death of a Normal Volunteer in Research," *Journal of Medical Ethics* 28, no. 1
(February 1, 2002): 3–4. https://doi.org/10.1136/jme.28.1.3.

14. Gina Kolata, "Johns Hopkins Admits Fault in Fatal Experiment," *New York
Times*, July 17, 2001.

15. Deborah Josefson, "Healthy Woman Dies in Research Experiment," *BMJ* 322
(June 30, 2001): 1565. https://www.bmj.com/content/322/7302/1565.2.

16. Savulescu and Spriggs, "Hexamethonium," 3.

17. Office for Human Research Protections (OHRP), "OHRP Letter to Hopkins on
Decision to Suspend Funding," *BaltimoreSun.com* (February 10, 2015). https://www
.baltimoresun.com/bal-ohrpletter-story.html.

18. Richard Ian Ogilvie, "The Death of a Volunteer Research Subject: Lessons to Be Learned," *CMAJ* 165 no. 10 (2001):1335–37.

19. Anne Tomlin, "Hospital Librarians and the Johns Hopkins Tragedy," *Journal of Hospital Librarianship* 2, no. 4 (September 1, 2002): 89–96. https://doi.org/10.1300/J186v02n04_07.

20. Robinson and Lipscomb Gehle, "Medical Research and the IRB," 21.

21. Laura L. Haines, Jeanene Light, Donna O'Malley, and Frances A. Delwiche, "Information-Seeking Behavior of Basic Science Researchers: Implications for Library Services," *Journal of the Medical Library Association* 98, no. 1 (January 2010): 73–81. https://doi.org/10.3163/1536-5050.98.1.019.

22. Melissa L. Rethlefsen, Shona Kirtley, Siw Waffenschmidt, Ana Patricia Ayala, David Moher, Matthew J. Page, Jonathan B. Koffel, and PRISMA-S Group, "PRISMA-S: An Extension to the PRISMA Statement for Reporting Literature Searches in Systematic Reviews," *Systematic Reviews* 10, no. 1 (2021): 39. https://doi.org/10.1186/s13643-020-01542-z.

23. CITI Program, "Collaborative Institutional Training Initiative," accessed January 15, 2021, https://about.citiprogram.org/en/homepage/.

BIBLIOGRAPHY

Cantwell, Laureen P., and Doris Van Kampen-Breit, "Librarians and the Institutional Review Board (IRB): Relationships Matter." *Collaborative Librarianship* 7, no. 2 (2015): 66–78.

Cheek, Fern M. "Research Support in an Academic Medical Center." *Medical Reference Services Quarterly* 29, no. 1 (January 2010): 37–46. https://doi.org/10.1080/02763860903485068.

CITI Program. "Collaborative Institutional Training Initiative." Accessed January 15, 2021. https://about.citiprogram.org/en/homepage/.

Fay, Brenda, Jennifer Deal, and Vicki Budzisz. "An Institutional Repository Experience at a Large Health Care System." *Medical Reference Services Quarterly* 36, no. 3 (July 3, 2017): 280–91. https://doi.org/10.1080/02763869.2017.1332264.

Federer, Lisa. "The Librarian as Research Informationist: A Case Study." *Journal of the Medical Library Association* 101, no. 4 (October 2013): 298–302. https://doi.org/10.3163/1536-5050.101.4.011.

Frumento, Katherine Stemmer, and Judith Keating. "The Role of the Hospital Librarian on an Institutional Review Board." *Journal of Hospital Librarianship* 7, no. 4 (December 2007): 113–20. https://doi.org/10.1300/J186v07n04_08.

Haines, Laura L., Jeanene Light, Donna O'Malley, and Frances A. Delwiche. "Information-Seeking Behavior of Basic Science Researchers: Implications for Library Services." *Journal of the Medical Library Association* 98, no. 1 (January 2010): 73–81. https://doi.org/10.3163/1536-5050.98.1.019.

Harvey, Sally. "Institutional Review Boards: Another Way for Hospital Librarians to Add Value to Their Organization." *Journal of Hospital Librarianship* 3, no. 2 (March 2003): 99–102. https://doi.org/10.1300/J186v03n02_09.

Josefson, Deborah. "Healthy Woman Dies in Research Experiment." *British Medical Journal* 322 (2001): 1565. https://www.bmj.com/content/322/7302/1565.2.

Kolata, Gina. "Johns Hopkins Admits Fault in Fatal Experiment." *New York Times*, July 17, 2001.

Labaree, Robert V. "Working Successfully with Your Institutional Review Board: Practical Advice for Academic Librarians." *College & Research Libraries News* 71, no. 4 (April 1, 2010): 190–93. https://crln.acrl.org/index.php/crlnews/article/view/8353/8484.

National Commission for the Protection of Human Subjects of Biomedical and Behavioral Research, Department of Health, Education and Welfare (DHEW). "The Belmont Report." Washington, DC: United States Government Printing Office, September 30, 1978. https://www.hhs.gov/ohrp/regulations-and-policy/belmont-report/read-the-belmont-report/index.html.

Office for Human Research Protections (OHRP). "OHRP Letter to Hopkins on Decision to Suspend Funding." *BaltimoreSun.com* (February 10, 2015). https://www.baltimoresun.com/bal-ohrpletter-story.html.

Ogilvie, Richard Ian. "The Death of a Volunteer Research Subject: Lessons to Be Learned." *Canadian Medical Association Journal* 165 no. 10 (2001):1335–37.

Raimondo, Paula G., Ryan L. Harris, Michele Nance, and Everly D. Brown. "Health Literacy and Consent Forms: Librarians Support Research on Human Subjects." *Journal of the Medical Library Association* 102, no. 1 (January 2014): 5–8. https://doi.org/10.3163/1536-5050.102.1.003.

Rethlefsen, Melissa L., Shona Kirtley, Siw Waffenschmidt, Ana Patricia Ayala, David Moher, Matthew J. Page, Jonathan B. Koffel, and PRISMA-S Group. "PRISMA-S: An Extension to the PRISMA Statement for Reporting Literature Searches in Systematic Reviews." *Systematic Reviews* 10, no. 1 (2021): Article 39. https://doi.org/10.1186/s13643-020-01542-z.

Robinson, Judith G., and Jessica Lipscomb Gehle. "Medical Research and the Institutional Review Board: The Librarian's Role in Human Subject Testing." *Reference Services Review* 33, no. 1 (January 1, 2005): 20–24. https://doi.org/10.1108/00907320410519360.

Savulescu, Julien, and Merle Spriggs. "The Hexamethonium Asthma Study and the Death of a Normal Volunteer in Research." *Journal of Medical Ethics* 28, no. 1 (February 1, 2002): 3–4. https://doi.org/10.1136/jme.28.1.3.

Smale, Maura A. "Demystifying the IRB: Human Subjects Research in Academic Libraries." *portal: Libraries and the Academy* 10, no. 3 (2010) (3): 309–21. https://doi.org/10.1353/pla.0.0114.

Tomlin, Anne. "Hospital Librarians and the Johns Hopkins Tragedy." *Journal of Hospital Librarianship* 2, no. 4 (September 1, 2002): 89–96. https://doi.org/10.1300/J186v02n04_07.

Visintini, Sarah, Mish Boutet, Alison Manley, and Melissa Helwig. "Research Support in Health Sciences Libraries: A Scoping Review." *Journal of the Canadian Health Libraries Association* 39, no. 2 (July 24, 2018): 56–78. https://doi.org/10.29173/jchla29366.

Part II

INSTITUTIONAL ANIMAL CARE AND USE COMMITTEE

Chapter Four

The Institutional
Animal Care and Use Committee

*Membership, Responsibilities,
and Roles for Librarians*

Susan M. Harnett

The roles of the health science library and health science librarians have evolved over the last two decades. Traditionally, the library is seen as an information warehouse, and the librarian as gatekeeper; however, the modern health science library has changed its mission in response to the needs of its community. In shifting the library's focus from print book and journals to electronic formats readily accessible to remote, affiliated patrons, creating collaborative spaces and emphasizing skill-based teaching, health science librarians have had the opportunity to redefine their jobs from inward-facing to outward, and explore new ways to provide value to the institution. Embedding into research demonstrates the value of librarians' skills and commitment to the strategic goals of the institution. Service on institutional regulatory committees such as the Institutional Animal Care and Use Committee (IACUC) is an opportunity for librarians to embed into the research lifecycle at the very beginning, as well as create partnerships within the institutional community.

The IACUC is established by two federal laws as the oversight authority for all research involving animal models: *The Public Health Service Policy on the Humane Care and Use of Laboratory Animals* (Public Law 99-158), under the regulatory agencies the National Institutes of Health (NIH) and the Office of Laboratory Animal Welfare (OLAW); and the *Animal Welfare Act* (9 CFR § 2.31), regulated by the United States Department of Agriculture (USDA).

As an oversight authority, the IACUC is charged with ensuring that institutions that use animals in research comply with all applicable laws and standards set forth by both agencies. Regardless of agency, the composition and duties of the IACUC are similarly mandated, with some variations in coverage; for example, the Animal Welfare Act (AWA) does not extend protections to birds, rats, and mice, but does extend protections to deceased warm blooded animals; while the Public Health Service (PHS) policy covers all live vertebrate animals. The PHS policy and the AWA often refer to each other in both descriptions of duties and enforcing regulations. An informative comparison table of both PHS and AWA requirements developed by Mildred Johnson, J.D., Creighton University, is available online at creighton.edu/fileadmin/user/ResearchCompliance/IACUC/Resources/comparison.PDF; information therein may be superseded by updates to these policies.

The minimum number of members on an IACUC is five for research funded by the Public Health Service and cooperating agencies, or three for research funded through the USDA. A five-member IACUC must include a chair of the committee; a veterinarian trained in use of research animals; an unaffiliated member who must not have any relationship to the institution or to a person affiliated with the institution; a scientist experienced in the use of research animals; and a nonscientist member. A minimum three-member IACUC must contain a chair, a veterinarian, and a non-affiliated member. In either case, the IACUC should strive to represent both the institution and surrounding community, as "diversification on the IACUC is intended to broaden the perspective and add depth to the . . . review processes."[1]

In theory, a librarian with appropriate education and experience may perform scientific roles and may also serve as chair. However, librarians generally serve the IACUC as unaffiliated or community members, or as nonscientists. Additionally, librarians may serve as ex-officio members, consultants, or liaison librarians embedded in clinical and bench science departments or into the committee itself.

THE UNAFFILIATED, COMMUNITY, OR LAY MEMBER

The unaffiliated member or community member must not be affiliated with the research institution in any capacity other than membership on the IACUC. Unaffiliated members (UMs) may also be referred to as community or lay members. Such members are recruited from outside the institution and should not have professional or personal relationships with other members prior to service. The UM may have a background in an area of research, though the unaffiliated member's role is to ensure that the committee consider ethical and humane implications of the proposed research rather than the scientific. As such, "the unaffiliated committee member should provide representation for general community interests in the proper care and treatment of animals."[2] The UM may be a professional in another discipline such as law or business, an academic in an institution other than the IACUC's home institution, a member of the clergy, or an interested community member of any profession.

Librarians may serve as UMs at an institution other than their own. For instance, an interested public or other academic librarian may serve as the UM on the IACUC of a local research university.

THE NONSCIENTIST

The nonscientist should not have any scientific background, including undergraduate or graduate degrees in the sciences. The nonscientist may be affiliated with the research institution in some capacity unrelated to scientific research. Responsibilities may vary among institutions, with some nonscientists required to review protocols, perform facility inspections, and investigate non-compliance issues. As with the lay member, a public or nonscientific academic librarian may serve in this capacity. PHS policy indicates that members may serve in more than one category; therefore, a librarian may serve as both the lay member and the nonscientist.

THE EX-OFFICIO MEMBER OR CONSULTANT

The IACUC may include an ex-officio member who serves by obligation or privilege of their position. Both PHS policy and the AWA indicate that voting rights are strictly limited to appointed members. Nevertheless, the IACUC may include ad hoc appointees whose special expertise in their disciplines is valuable to the board. It is up to each individual IACUC as to whether and how consultants are utilized.

A librarian's expertise in literature searching and information literacy, for example, may prove valuable to investigators as they complete the federally mandated requirements of the literature search and lay language sections of the protocol. A consultant librarian may also attend regular meetings of the board to provide real-time clarification of issues raised during the meeting. As an example, a consultant librarian, after reviewing a protocol, noticed that the mode of administration for a listed analgesic was not recommended for a particular strain of rat. She was able to inform the board of this discrepancy, leading to a modification of the protocol. Had the protocol passed as written, the improperly administered analgesic may have resulted in prolonged pain and suffering of the study animals and potentially may have affected the study's outcome.

Librarians skilled in areas such as bioinformatics, biostatistics, data management, and analysis may serve to inform investigators as they develop animal protocols and analyze results. Additionally, a librarian with such expertise may assist the board in assessing the methodology of the protocol, make recommendations to investigators for data management and analysis, or assist in grant writing.

THE LIAISON OR EMBEDDED LIBRARIAN

Librarians at large research institutions with liaison programs may be embedded within the IACUC and/or research programs using animal models in accordance with library and institutional policies. In that capacity, they may train investigators in performing the literature search required by the Animal Welfare Act. They may perform consultations for new protocols. In some institutions, new investigators are required to have both a consultation and training session with the librarian prior to submitting any protocol. The embedded or liaison librarian may review the required literature searches and provide feedback as to adequacy of the search and its findings, or update searches for renewals. They may review the lay language section of the protocol for comprehension, jargon, and terminology. As IACUC searches can be complex and time-consuming, liaison librarians may opt not to attend meetings or review all submissions, instead focusing on individual requests for assistance from the committee or investigators. Liaison librarians can additionally connect the IACUC and investigators to sources of expertise on campus, as they often collaborate with a variety of different departments and faculty.

THE SCIENTIST MEMBER

Librarians with the appropriate backgrounds may also serve as scientific members of the IACUC. As such, they are charged not only with evaluating the protocol for humane concerns and compliance, but also with assessing the scientific methodology, justification of the number of animals to be used, and statistical validity of the proposed research. It is rare, but not unheard of, for a librarian to serve as chair of the committee.

Ideally, IACUC members should adequately represent both the research institution's scientific and surrounding communities. This includes scientist members who have expertise in the use of laboratory animals but may also include those whose research is primarily clinical. Biostatisticians, legal counsel, institutional Occupational Safety and Health Administration (OSHA) representatives, librarians, and others who may be involved in the research lifecycle lend valuable expertise to the work of the IACUC. Community, nonscientist and lay members represent the general public and their understanding of the proposed animal research and lend transparency to the process. The non-scientist and unaffiliated members provide, ". . . a clear and concise sequential description of the procedures for the use of animals that is easily understood by all members of the committee."[3] Though nonscientist and unaffiliated members are not expected nor required to have scientific knowledge, it is helpful to have some awareness of biomedical research and its methods, the ethics of research, intellectual curiosity and confidence, and the ability to understand and translate scientific and legal concepts to the community.[4]

PREPARING TO SERVE AS AN IACUC MEMBER: TRAINING AND CERTIFICATIONS

Once appointed to the IACUC, the prospective member must undergo training and certification as required by the USDA and OLAW. Basic courses are "Working with the IACUC" and "Essentials for IACUC members," which are available through the Collaborative Institutional Training Initiative (CITI; https://about.citiprogram.org/en/homepage). Members must renew these certifications every three years. There are additional, optional courses, which may be assigned as needed, on the roles of IACUC members, and more specific topics in animal use. These courses cover the composition of the committee, the responsibilities of members, federal laws, policies and guidelines, humane care and handling of animals, protocol review, and non-compliances. IACUC members may opt for additional training from other organizations. A comprehensive list of training opportunities is available from the American Association for Laboratory Animal Science (AALAS; https://www.aalas.org/iacuc/training-resources/iacuc-members-staff). The IACUC may periodically offer additional training for its members in the form of workshops and retreats, and sponsor professional development opportunities such as the IACUC 101 Series (OLAW).

For librarians who may be performing or reviewing literature searches, it may be helpful to attend an alternatives literature searching workshop, such as "Meeting the Information Requirements of the Animal Welfare Act" at the Animal Welfare Information Center (AWIC). Sponsored by the USDA and National Agricultural Library, this workshop concentrates on the federally mandated literature search for alternatives to animal use and procedures that require more than momentary pain and distress.

WHAT ARE THE RESPONSIBILITIES OF AN IACUC MEMBER?

Depending on the institution, non-affiliates, community members, and non-scientists may or may not review protocols. Reviewers are generally assigned protocols a few weeks before the scheduled meeting of the full committee. Most institutional IACUCs use an online, secure reviewing system that allows the reviewer to see the protocol and note any changes for the investigator. The AWA (9 CFR §2.31, d) lists the conditions that must be met before any protocol involving the use of animals is approved. These include the following:

- Procedures will avoid or minimize discomfort, stress, and pain in the animal.
- The principal investigator has considered alternatives to procedures that cause more than momentary pain or distress and has provided a written narrative that includes methods undertaken to determine whether these alternatives are available.
- The proposed procedure does not unnecessarily duplicate previous experiments.
- For those procedures that may cause more than momentary distress, the investigator will provide appropriate anesthesia and analgesia. Investigators must justify scientifically, in writing, any procedure in which anesthesia and analgesia cannot be administered.
- Animals for whom pain and distress cannot be relieved, or whose endpoint is terminal, will be humanely and painlessly euthanized.
- Medical care will be provided by veterinary staff as needed.
- All surgeries will provide for pre- and post-operative care in accordance with established veterinary principles, including aseptic procedures in both dedicated and non-dedicated facilities.
- The animals will be housed appropriately for their species, and provided food, water, bedding, and enrichment.

Reviewers may also look for the experience, education, and certification of the primary and co-investigators; their qualification to perform the proposed procedures; and the supervision of lab personnel. Investigators should clearly state, in a lay summary, the rationale for the research, its methods, and its potential for both harm and benefit. They should describe all procedures, anesthesia, analgesia, response to adverse events, and humane endpoints. The investigator is required by law (9 CFR §2.31 (d)(ii-iii)) to perform a literature search in at least two separate databases, addressing alternatives to the use of animals in the protocol, reduction in the number of animals used, and minimization of pain and distress for the animal. The investigator should indicate the species, genus, and source of the animals and their disposition, whether euthanasia or transfer to another protocol, after the protocol has closed; he or she should be able to justify animal numbers requested and indicate pain categories, B through E (with B assigned to animals being held or bred for a protocol or observation only and E for more than momentary pain or distress that is unrelieved by anesthesia or analgesia). The IACUC will decide as to the appropriate pain and distress category at the time of review.

Reviewers present the protocol at a regularly scheduled meeting of the full committee for discussion. Investigators may opt to be present during the discussion to answer any questions from the committee. Exceptions to this are protocols that are assigned to Designated Member Review. These protocols are generally those in which category B and C animals are to be used, in which the animals will be subjected to only momentary distress or pain, behavioral/observational studies, or breeding programs. Category D and E studies are always presented to the full committee as the investigator is required to justify any procedure causing more than momentary pain to the experimental animal. Continuing reviews, renewals, and amendments to protocols are also brought to the full committee if animals are used in study categories D or E.

Members may also be required to participate in semi-annual facility inspections. This is an activity required by both PHS Policy and the AWA, and is described in the *Guide for the Care and Use of Laboratory Animals*.[5] Facility inspections are reportable to the Institutional Official (IO). Additional reporting may be requested by OLAW and the Animal and Plant Health Inspection Service (APHIS), a subagency of USDA. Any area in which an animal lives or breeds, or where food, medication, and equipment relating to animal use are stored, as well as all approved procedure rooms, must be inspected. USDA animal facilities must be inspected by a minimum of two IACUC members; non-USDA facilities may have just one member perform the inspection. A representative of the IACUC administrative staff may accompany the IACUC members. The team may be joined by representatives of Animal Care Services and/or veterinary staff, as well as the primary investigator or a team member. All questions should be directed to laboratory staff, who must be able to explain how animals are used in that particular space. Many institutional IACUCs have developed their own checklists for use during inspections. Here are some examples of relevant questions:

- What is the primary use for this room? (Housing, procedure, storage, etc.)
- What happens to animals in this room?
- Where do the animals come from? Do they live in this space, or are they brought from another area? How are they transported?

- How much time do they spend here?
- How is the procedure performed? Is it a survival or non-survival surgery? How are the animals anesthetized? How is their pain controlled? Where do they recover? How many procedures does the staff perform in a day? Are there designated areas for prep, surgery, and recovery?
- How is the sanitary field maintained? Do investigators wear appropriate protective clothing while handling animals or performing procedures?
- What form of anesthesia is used? How is it administered? How do you ensure that the animal is completely anesthetized before beginning the procedure?
- Is a copy of the protocol available to all lab members? Have all lab members read the protocol?

Inspectors may observe, in accordance with institutional policies, that areas are specifically designated as preoperative, procedural, and postoperative (recovery); that there is overall cleanliness, orderliness, and good repair of the area; and that there are adequate, dated, and labeled cleaning supplies that are stored appropriately. Animal food and treats must be clearly labeled and dated; and stored in rodent-proof containers with tight-fitting lids or if unopened, on pallets off the floor. The inspector notes whether animals have appropriate enrichment opportunities, which may range from mouse-specific chewing materials to outdoor activities and toys for other species. All food, water, and bedding changes must be noted and dated. Inspectors must check acquisition, sanitation, and expiration dates of all products or drugs that are used in the protocol. If applicable, investigators must present Drug Enforcement Agency (DEA) licensing for all controlled substances, and these must be in a double-locked immovable location, with uses recorded in a log book kept with the substances. Expired medications must be clearly labeled as such, segregated from current medications, and disposed of in accordance with university and IACUC policy.

Any discrepancies or deficiencies must be noted and be designated as a minor or major deficiency, which may result in a request from the IACUC for remediation within a certain time frame. An example of a minor deficiency would be an expired unopened bag of animal feed. This may be remedied by disposal under the inspector's direction. Expired medications that continue to be administered to subject animals may be considered a major deficiency. Major deficiencies, if not remedied in a satisfactory manner, may result in a non-compliance investigation. Repeated non-compliances may result in the suspension of the investigator's protocol or banning of the investigator from animal work for a period of time. The IACUC will vote to accept or reject these discrepancy reports at the regularly scheduled meeting and may recommend further reparative action, including mandated reporting to the appropriate regulatory agency.

Facilities may be on-site, as in animal housing or procedure rooms located in campus buildings; or off-site, as in farms, barns, stables, or pens. Each may require sterile procedures. Gowns, gloves, and shoe covers may be required for certain rooms, while full head-to-toe contamination suits, respirators, and hoods may be required for others. Though most investigators make every effort to avoid procedures during the inspection, IACUC members may view surgical or euthanasia procedures, which may be distressing.

Prior to entering facilities, IACUC members must complete health and risk assessments as required by the institution. Additional vaccinations and assessments may be required to enter certain facilities. This is to protect the animals from any human acquired disease, not necessarily to protect the inspector from the animal; though not touching animals, their habitats, food, or feces is considered a best practice to maintain the sanitary field. It is important to follow any directions from the veterinary staff and investigators as inspections can be disruptive to protocols involving nocturnal animals and/or sleep-wake cycles. Inspectors may also perform inspections via webcam in locations that may not be readily accessible to IACUC members. The chapter author, as a nonscientist member, has inspected black rhino pens, whale necropsy suites, a bedbug research facility, horse treadmills, and a surgical suite during a spay and neuter training for veterinary students.

Members may also be asked to serve on various committees, such as those investigating non-compliance issues. A non-compliance occurs when investigators or staff deviate from the approved protocol. These deviations may be accidental, or they may be more deliberate. Most non-compliances are the result of staff training deficiencies and communication failures. Investigators must supervise their teams and ensure that the description of animal care and use is followed exactly as written into the protocol; any deviation may result in a non-compliance. A non-compliance can be self-reported, as in the investigator realizes a mistake has been made and alerts the IACUC. Non-compliances can also be reported anonymously, and contact information for whistle-blowers should be posted in every animal area. The IACUC is charged with investigating every occurrence of a possible non-compliance, reporting it to the university officer and any applicable agency. The IACUC may seek rectification of the circumstances that led to the non-compliance; this may include retraining and additional supervision of the employee; an evaluation of lab policies; or reassignment of investigators and staff. For particularly egregious or repeated offenses, the IACUC has the authority to halt all animal work on the protocol and can suspend *all* of the investigator's protocols, which may result in the loss of time, prestige, and funding for the investigator. It may also result in fines for the institution. It is, therefore, in the institution's best interest to support the IACUC in its mission of compliance with all federal, state, and local laws, as well as university policies governing the use of animals in research.

Additionally, the use of animals in research is a hotly debated, emotional issue for many in the community. For this reason, the names of committee members are rarely published on the institutional website, and in some institutions, meetings are closed-door. In states with open meeting laws, committee meetings are open to the public. Though a rare occurrence, members should be aware that service on the IACUC may provoke negative reactions from the community. For nonscientists and UMs unfamiliar with the use of animal models in research, it can be challenging to reconcile the ethical dichotomy of serving on the IACUC while advocating for animal welfare.

WHY SHOULD LIBRARIANS SERVE ON THE IACUC?

Service on the IACUC is a high impact activity that allows librarians to actively connect with faculty and researchers who may rarely step into the physical library or who are outside of the liaison area. By meeting researchers where they are, the librarian has the opportunity not only to market existing services, but develop new ones based on researcher needs. The OCLC Research "Call to Action" argues that academic libraries must grow services to reflect the changing needs of the research environment, including

- designing services around the research process,
- embedding and integrating librarians and content within the researcher workflow,
- assessing and training staff to support new modes of research,
- making content discoverable by researchers,
- seeking collaborative partnerships, and
- assessing library services and resources to demonstrate value. (OCLC Research, 2009)[6]

A librarian who participates as a non-scientist or community member also brings necessary diversity to the committee. Research indicates that an IACUC whose membership is too heavily skewed toward animal researchers may result in a committee in which emphasis is directed to experimental methodology rather than the ethics of animal use in research.[7] Additionally, an IACUC in which the majority of members are animal researchers may suffer from approval bias and groupthink in evaluating protocols, as "in likeminded groups, the opinions sought and shared are those that reinforce, rather than challenge the favoured conclusion."[8] An IACUC with a more diverse membership may encourage diversity of opinion, as the "mere awareness of a dissenting voice in a group is enough to alter cognitive processes and yield better judgments" (Hansen 2012).[9] The inclusion of community members and nonscientists also helps to lessen what may be perceived as an intimidating atmosphere in a heavily scientific committee, as the successful IACUC "requires active, informed participation of all its members and their values and knowledge."[10]

A literature search for alternatives is federally mandated for all research involving covered species. The Code of Federal Regulations' content on Animals and Animal Products (9 CFR, Part 2, Sec. 2.31 (d)(1)(ii)) requires that the " IACUC shall determine that . . . the principal investigator has considered alternatives to procedures that may cause more than momentary or slight pain or distress to the animals, and has provided a written narrative description of the methods and sources, e.g., the Animal Welfare Information Center, used to determine that alternatives were not available."[11] The Federal Register (1989) further elaborates that the "principal investigator must provide a written narrative of the sources, such as biological abstracts, MEDLINE, the Current Research Information Service (CRIS), and the Animal Welfare Information Center that is operated by the National Agricultural Library."[12] Protocols are additionally required to contain a lay language section describing the experiment as if written for a general interest publication. Therefore, the librarian's expertise in literature searching, database selection, resource identification, and information literacy is a valuable asset to the work of the IACUC.

Librarians are poised to stake a claim in research—not just as an "information-gatekeeper," but as collaborative partners, data managers, bioinformationists, and other emerging roles. Health science librarians, in particular, have experienced a transformation in job functions, from research support services to inclusion as members of scientific research teams. "Many librarians have advanced training in both subject-matter disciplines and information science. It is common to find librarians working as part of health care teams, writing grant proposals, serving on institutional review committees, working as bioinformatics database specialists within science departments, [and] serving as faculty members in evidence based medicine courses."[13] The experience of working with biomedical researchers confers a deep understanding of the research process, though not necessarily the science behind the research.[14] Service to the IACUC allows librarians to impact research at the start of, and throughout, the research lifecycle.

NOTES

1. Christian E. Newcomer and William G. Greer. "General Composition of the IACUC and Specific Roles of the IACUC Members," In The *IACUC Handbook*, eds.

Jerald Silverman, Mark A. Suckow and Sreekant Murthy (Boca Raton: CRC Press, Taylor & Francis Group, 2014), 45.

2. Rebecca Dresser. "Community representatives and nonscientists on the IACUC: What difference should it make?" *ILAR Journal* 40, no. 1 (1999): 29.

3. Christian E. Newcomer and William G. Greer. "General Composition of the IACUC and Specific Roles of the IACUC Members," In *The IACUC Handbook*, eds. Jerald Silverman, Mark A. Suckow and Sreekant Murthy (Boca Raton: CRC Press, Taylor & Francis Group, 2014), 66.

4. Joan P. Porter, "What are the ideal characteristics of unaffiliated/nonscientist IRB members?" *IRB ethics & Human Research* 8, no. 3 (1986): 3–4.

5. Committee for the Update of the Guide for the Care and Use of Laboratory Animals, Institute for Laboratory Animal Research, Division on Earth and Life Studies, *Guide for the care and use of laboratory animals.* (Washington, DC: National Academies Press, 2014).

6. Chris Bourg, Ross Coleman, and Ricky Erway, "Support for the research process: An academic library manifesto," Report produced by OCLC Research, November 2009, accessed September 10, 2021, http:// www.oclc.org/content/dam/research/publications/library/2009/2009-07.pdf.

7. Lawrence Arthur Hansen, "Institutional animal care and use committees need greater ethical diversity," *Journal of Medical Ethics*, 39 (2013): 188–190.

8. Ibid.

9. Ibid.

10. Jerald Silverman, et.al., "Factors influencing IACUC decision making: Who leads the discussion?" *Journal of Empirical Research Human Research Ethics* 12, no. 4 (October 2017): 214.

11. *"Institutional Animal Care and Use Committee," Title 9 Code of Federal Regulations*, Pt. 2, 2020. Accessed September 10, 2021, https://www.govinfo.gov/app/details/CFR-2020-title9-vol1/CFR-2020-title9-vol1-chapI-subchapA.

12. Department of Agriculture, Final Report. "9 *CFR Parts 2 and 3-Animal Welfare". Federal Register* 54 no. 168 (31 August 1989):36130. Accessed September 10, 2021, https://www.govinfo.gov/content/pkg/FR-1989-08-31/pdf/FR-1989-08-31.pdf

13. Donald A.B. Lindberg and Betsy L. Humphreys, "2015-Future of Medical Libraries," *New England Journal of Medicine* 352 (2005): 1067–70.

14. Emily Glenn and Betsy Rolland, "Librarians in Biomedical Research: New Roles and Opportunities," *Information Outlook* 14, no.7 (2010): 26–29.

BIBLIOGRAPHY

"Animal Welfare Act," 7 U.S.C. 2131-2160, 9 CFR § 2.31 (1966).

Animal Welfare Information Center (AWIC). "About AWIC." https://www.nal.usda.gov/awic/about-awic.

Bourg, Chris, Ross Coleman, and Ricky Erway. "Support for the Research Process: An Academic Library Manifesto." *OCLC Research* (2009). https://www.oclc.org/content/dam/research/publications/library/2009/2009-07.pdf.

Committee for the Update of the Guide for the Care and Use of Laboratory Animals, Institute for Laboratory Animal Research, Division on Earth and Life Studies. *Guide for the Care and Use of Laboratory Animals.* Washington, DC: National Academies Press, 2011.

Department of Agriculture. Final Report. "9 CFR Parts 2 and 3-Animal Welfare." Federal Register 54, no. 168 (August 31, 1989): 36130–36163. https://www.gov info.gov/content/pkg/FR-1989-08-31/pdf/FR-1989-08-31.pdf.

Dresser, Rebecca. "Community Representatives and Nonscientists on the IACUC: What Difference Should It Make?" *ILAR Journal* 40, no. 1 (1999): 29–33. https:// doi.org/10.1093/ilar.40.1.29.

Glenn, Emily, and Betsy Rolland. "Librarians in Biomedical Research: New Roles and Opportunities." *Information Outlook* 14, no. 7 (2010): 26–29.

Hansen, Lawrence Arthur. "Institutional Animal Care and Use Committees Need Greater Ethical Diversity." *Journal of Medical Ethics*, 39, no. 3 (2012): 188–90. doi:10.1136/medethics-2012-100982.

Institutional Animal Care and Use Committee (IACUC), 9 C.F.R §2.31(d) (2004).

Johnson, Mildred. "Animal Research Requirements Comparison Table," As of October 11, 2001. Accessed September 15, 2021. creighton.edu/fileadmin/user/ ResearchCompliance/IACUC/Resources?comparison.PDF.

Lindberg, Donald A. B., and Betsy L. Humphreys. "2015—The Future of Medical Libraries." *New England Journal of Medicine* 352 (2005): 1067–70. DOI: 10.1056/ NEJMp048190.

Porter, Joan P. "What Are the Ideal Characteristics of Unaffiliated/Nonscientist IRB Members?" *IRB Ethics & Human Research* 8, no. 3 (1986): 1–6.

Silverman, Jerald, Charles W. Lidz, Jonathan Clayfield, Alexandra Murray, Lorna J. Simon, and Louise Maranda. "Factors Influencing IACUC Decision Making: Who Leads the Discussion?" *Journal of Empirical Research Human Research Ethics* 12, no. 4 (2014): 209–16. doi:10.1177/1556264617717827.

Silverman, Jerald, Mark A. Suckow, and Sreekant Murthy (Eds.). *The IACUC Handbook.* Boca Raton, FL: CRC Press, 2014.

U.S. Department of Health and Human Services (DHHS), National Institutes of Health (NIH), Office of Laboratory Animal Welfare (OLAW). *Public Health Service Policy on Humane Care and Use of Laboratory Animals.* Retrieved from https://OLAW.nih.gov.

USDA Animal Care: Animal Welfare Act and Animal Welfare Regulations. Washington, DC: United States Department of Agriculture (USDA), Animal and Plant Inspection Service (APIS), 2017.

Chapter Five

A Brief History of Animal Research in the United States and Librarian Involvement in the IACUCs

Karen D. Barton

Major advances in biomedical science such as successful organ transplantation, the germ theory that microorganisms can cause certain diseases, and the recent, groundbreaking discovery of the hepatitis C virus are due largely in part to animal research.[1] Studying animal models of diseases has had both clinical and public health benefits, improving quality of life and life expectancy in various populations around the world. In fact, of the 222 Nobel laureates in the physiology or medicine category, 186 used animal models in their research.[2] As rodents make up 95 percent of animals needed for biomedical research, studying disease models across the life span and multiple generations of animals with shorter life cycles than humans is critical to understanding disease processes in humans more quickly.[3] Throughout centuries, the field of laboratory animal science is one that has had to evolve, however, to adhere to legal and ethical standards for the care and use of animals. The judgment of respected thought leaders, persistent lobbying of animal welfare organizations, and public outcry have been significant catalysts for this change.

Vertebrate animals have been used by humans as models for their anatomy and physiology since the dawn of medicine, and evidence exists of dissection and vivisection—experimental surgery on live animals—in ancient Greece.[4] Various physiological experiments and differing beliefs of human dominion over animals as well as differing beliefs regarding the ability of animals to feel pain continued for centuries.[5] As the amount of animal research increased in the second half of the nineteenth century in certain parts of the world, opposition to vivisection gained momentum, which led to the creation of animal welfare and animal rights organizations and movements. The public debate surrounding vivisection was partly fueled by Charles Darwin's publication of *On the Origin of Species* in 1859, which discussed similarities between humans and animals and gave both scientists and antivivisectionists an argument for their cause.[6] As a result of antivivisectionist activities in Great Britain, the Cruelty to Animals Act of 1876 was passed by the British.[7] It required licensing and facility inspections and placed some restrictions on painful experiments.

The Cruelty to Animals Act of 1876 greatly influenced activism and policy in the United States. Less than a decade after the act passed, and while no federal legislation restricting animal research had yet been passed in the United States, an American Medical Association (AMA) committee was formed to oppose such legislation. That legislation included the "Gallinger-DC" bill, which was modeled after and may have been more restrictive than the British Cruelty to Animals Act. The bill, which could have set a precedent for laws in various states, was an effort to regulate animal research in the District of Columbia and it posed many threats to biomedical research. Those threats included its likelihood to prohibit very basic research and the fact that commissioners with no scientific knowledge could be appointed in order to determine the potential of the research to advance the discipline of physiology and to determine if the research met other criteria. Between 1896 and 1900, antivivisectionists fought to have the Gallinger-DC bill passed, yet they were not successful.[8]

Harvard University and the Johns Hopkins University established animal facilities in the 1880s, becoming the first U.S. institutions to do so.[9] Later, in 1904, Dr. Milton Rosenau, the director of the Hygienic Laboratory in Washington, DC, stated, "Animals are to be used in the proper work of the lab, but anything which inflicts pain upon them will not under any circumstances be allowed."[10] The Hygienic Laboratory was renamed the Public Health Service (PHS) in 1912, which included what is now known as the National Institutes of Health. In 1915, the Mayo Clinic became the first institution to have an animal facility managed by a veterinarian.[11]

In the early to middle part of the twentieth century, insufficient product safety testing as well as the Nuremburg Code, an ethics declaration in response to involuntary human experimentation in a Nazi program, were two main factors that contributed to the increasing amount of animal research being conducted in the United States.[12] During World War II, however, advancements in biology slowed.[13] After the war, "Rules Regarding Animals" were issued by the NIH director in 1950, then later revised as part of the NIH manual.[14] Also in 1950, the Animal Care Panel (ACP)—now the American Association for Laboratory Animal Science (AALAS)—was formed by a group of veterinarians.[15] At that meeting, the ACP's first president, Nathan Brewer, stated, "This organization is the logical outcome of the realization by research laboratories that their animal colonies are becoming increasingly the responsibility of each institution, rather that of each researcher."[16] In 1963, the ACP went on to publish the *Guide for Laboratory Animal Facilities and Care*—titled *Guide for the Care and Use of Laboratory Animals* in later editions. Based on scientific research and expert opinion, the guide is intended to assist researchers with developing an animal care and use program and provides detailed recommendations on topics such as personnel management and training, animal procurement, housing, and veterinary care.

Prior to the publication of the guide, the book *The Principles of Humane Experimental Technique* (1959) by British scientists W. M. S. Russell and R. L. Burch introduced "The Three R's"—reduction, refinement, and replacement—as an ethical approach to using animals for scientific research. As stated by Russell and Burch,

> Replacement means the substitution for conscious living higher animals of insentient material. Reduction means reduction in the numbers of animals used to obtain information of given amount and precision. Refinement means any decrease in the incidence or severity of inhumane procedures applied to those animals which still have to be used.[17]

At the beginning of each of the book's chapters, Russell and Burch quote Charles Darwin, whom they credit as being highly concerned about the welfare of experimental animals and responsible for furthering humane experimental technique.[18]

Also prior to the release of the guide, the Institute of Animal Resources (IAR)—now known as the Institute for Laboratory Animal Research (ILAR)—had been established as part of the National Research Council within the National Academy of Sciences to compile information on sources of research animals. ILAR has become instrumental in providing guidance related to all aspects of animal care and use to the international biomedical community in the form of reports, online resources, and programs. Its recommendations have influenced national and international laws and policies.[19] The 1950s also saw the founding of the American College of Laboratory Animal Medicine (ACLAM), which was first established as the American Board of Laboratory Animal Medicine (ABLAM). ACLAM is a training and certifying organization for veterinarians specializing in laboratory animal medicine. Dr. Nathan R. Brewer, who was influential in founding the ACP, ILAR, and ACLAM, served as the organization's first president.[20]

Today, three federal agencies oversee animal welfare in research: the USDA's Animal and Plant Health Inspection Service (APHIS), the National Institutes of Health (NIH), and the Food and Drug Administration (FDA). After the 1950s, the United States saw a major growth in efforts by federal agencies to provide animal welfare oversight. In 1965, a female Dalmatian named Pepper was stolen from a Pennsylvania family's farm, and it was soon discovered that the dog had been implanted with an experimental pacemaker in a research hospital facility in the Bronx and then cremated after she died due to the procedure.[21] At the time, animal welfare groups had been warning the public for years that pets were being snatched and sold to laboratories for experiments. Pepper's case made national headlines and the following year, Congress passed the landmark Laboratory Animal Welfare Act (PL 89-554) in order to regulate the transportation, sale, and handling of animals to be used for research and prevent pets from being stolen and sold to research laboratories.

Initially limited to covering dogs, cats, nonhuman primates, rabbits, hamsters, and guinea pigs, the Laboratory Animal Welfare Act was amended to include more species and renamed the Animal Welfare Act (AWA) in 1970 and then amended several times thereafter. In contrast to biomedical research that studies animals for the benefit of humans, wildlife research is typically conducted to benefit the animals under study. Field studies conducted on wildlife in their natural habitat and that are also not invasive, harmful, or behavior-altering are exempted from the AWA and IACUC oversight. The USDA became the responsible agency for enforcing the act and later established APHIS to further enforce it. APHIS may make unannounced inspections of research facilities and provides education and training on humane animal care.

The USDA, NIH, and FDA all require that institutions carrying out federally funded animal research have an Institutional Animal Care and Use Committee (IACUC) to oversee the care and use of animals at registered institutions, including approving or suspending animal research and conducting facility inspections. Additionally, these agencies have a memorandum of understanding for reciprocal cooperation. In 1971, the NIH Policy, Care, and Treatment of Laboratory Animals became the first "policy" applicable to institutions receiving NIH grant or contract awards; though during much of the 1970s, institutions could demonstrate compliance with NIH policies by being accredited by AAALAC International.[22]

For the regulation of animal testing for FDA-regulated product development, the Good Laboratory Practice for Nonclinical Laboratory Studies became law in 1979 and is enforced by the FDA's Office of Regulatory Affairs. The FDA also advocates for and adheres to the AWA and federal policies. In 1983, the U.S. Interagency Research Animal Committee (IRAC) formed and soon thereafter released the *U.S. Government Principles for the Utilization and Care of Vertebrate Animals Used in Testing, Research and Training* in 1985. Government agencies that perform or sponsor procedures for testing, research, or training involving vertebrate animals must adhere to the nine principles and are referred to the *Guide for the Care and Use of Laboratory Animals* for further guidance regarding their use.

The Health Research Extension Act of 1985 provided the statutory mandate for the PHS Policy on Humane Care and Use of Laboratory Animals, which requires the establishment of IACUCs in institutions that conduct federally funded research. It requires that NIH and other national research institutes ensure that awardees of their grants and contracts for research involving animals utilize the *Guide for the Care and Use of Laboratory Animals* and have an Animal Welfare Assurance. An assurance is a contract with the federal government that describes the institution's animal use and care programs and documents institutional commitment to follow all applicable laws and regulations.[23] Assurances must be in accordance with the PHS Policy on Humane Care and Use of Laboratory Animals and be approved by the NIH's Office of Laboratory Animal Welfare (OLAW), which monitors compliance of funded institutions in part by requiring annual reports and conducting site visits.

LITERATURE SEARCHING FOR
ANIMAL ALTERNATIVES AND IACUC MEMBERSHIP

A month after the Health Research Extension Act of 1985 called for the creation of animal care committees at entities that receive federal funding, the 1985 amendment to the Animal Welfare Act established an information service within the USDA's National Agricultural Library and in cooperation with the National Library of Medicine (NLM). That information service became the Animal Welfare Information Center (AWIC) in 1986. As stated in the AWA, the purpose of AWIC is to provide information that could assist with personnel training and prevent duplication of animal experimentation. It is also intended to be used to find information on improving methods of animal experimentation that could potentially reduce or replace animal use and minimize pain and distress to animals. The act essentially enforces the use of the 3Rs as defined by Russell and Burch and it mandates that IACUCs enforce its provisions to minimize pain and distress to animals.

Though there are no specific, standard requirements for searching for animal alternatives, it is the responsibility of the principal investigator to make "a good faith effort to demonstrate whether or not alternatives exist and why he/she will or will not adopt them."[24] AWIC suggests a literature search as the best way to demonstrate this and strongly encourages the use of multiple biomedical databases. Additionally, Title 9 of the *Code of Federal Regulations* (*9CFR*, Part 2, Sec. 2.31 (d)(1)(ii)) mentions that researchers should provide a written description of the methods and sources, with AWIC listed as an example source. While veterinary medical libraries, the National Agricultural Library, and the Medical Library Association were established as early as the nineteenth century, federal requirements to search for animal alternatives have only fueled librarian involvement in animal research, including their service as members of the IACUC.

The Health Research Extension Act of 1985 outlines that IACUCs must consist of a minimum of three members, including at least one member that has no association with the institution and at least one member that is a veterinarian. In addition to mandating that animal alternatives must be sought by principal investigators, the 1985 amendment to the AWA expands on how IACUCs should operate in regards to meeting and reporting. At each institution, the specific ways that IACUCs operate may vary. Some librarians who assist with conducting searches for animal alternatives are also members of the IACUC. As part of committees, librarians may participate in such activities as voting on animal use protocols, conducting research facility inspections, and reviewing research protocol amendments.

SUPPORT FOR IACUC LIBRARIANS

As previously mentioned, AWIC has been a significant source of support for IACUC librarians as its information specialists teach and provide resources online and in person for searches for animal alternatives. AWIC also assists librarians in understanding the AWA, USDA requirements for IACUCs and animal researchers, and provides information on other federal, state, local, and international laws and guidelines. Additionally, the Medical Library Association (MLA) has provided support systems in the form of interest groups for librarians who carry out work related to animal research.

In 1974, the MLA Veterinary Medical Libraries Group—later renamed the Veterinary Medical Libraries Section or VMLS—held its first business meeting a year after a dozen librarians held an unofficial meeting at the 1973 MLA annual meeting.[25] Formed in the decade prior to the passing of the AWA amendment that mandated the creation of IACUCs, the group did not have a specific focus of supporting the IACUC librarians. As stated in a 2008 proposal for an MLA special interest group (SIG) by Melissa Ratajeski, assistant director for data and publishing services at the University of Pittsburgh Health Sciences Library System, "Although there is a Veterinary section within MLA, many librarians supporting the IACUC are from large academic libraries and do not belong to the Veterinary section. In addition, IACUC searches often require extensive literature reviews on areas of research [such] as transplantation, AIDS, and tuberculosis which often falls out of the Veterinary section's scope."[26] SIGs were ad hoc groups open to all members of the association and did not require formalities such as elections for leadership positions as sections did.

Ratajeski, who had joined MLA the year prior to proposing the IACUC SIG, identified that there was a need for an organization or group within MLA "where librarians can turn to for help with such [IACUC] responsibilities, whether for advice, suggested resources, or sample search strategies." Feedback from MLA members to Ratajeski for the initial proposal for a SIG for IACUC librarians showed that some MLA members were in favor of a joint institutional review board (IRB) and IACUC SIG; however, it was determined that a group focused solely on the IACUC would be in their best interest. In December 2008, MLA approved the formation of the Institutional Animal Care and Use (IACU) SIG, which first convened at the 2009 MLA annual meeting with Ratajeski leading the group.[27]

The final IACU SIG proposal's stated purpose and goals were to allow MLA members to exchange knowledge, foster communication and collaboration, and act as a resource to MLA members who conduct animal alternative literature searches, support their institution's IACUC in any capacity, or have an interest in animal alternatives. By August 2009, the IACU SIG was comprised of forty-eight members and they have been successful in developing relevant conference programming and professional development opportunities.[28] The IACU SIG and Veterinary Medical Libraries Section both became caucuses and then subsequently merged to become the Animal and Veterinary Information Specialist (AVIS) Caucus in 2020. In December 2020, the AVIS Caucus had over 120 members.

IN THE PRESENT

Throughout history, multiple U.S. agencies have worked together for the same purpose of enforcing the ethical and humane treatment of animals and requiring justification for their use in research. This has largely been done by the passing of legislation that has introduced standards and best practices for institutions to follow as well as through requirements for regular reporting and auditing that help ensure that the standards are met. In turn, institutions can feel more assured that their transparency and compliance fulfills a responsibility to society, particularly since animal research is often conducted using taxpayer dollars and considering that many members of the public as well as some organizations continue to oppose the use of animals in research.

While investigators have legislation, literature, and a wide network of colleagues in which to find guidance, librarians have largely relied on each other for information and support in assisting institutions with meeting federal and other requirements. Currently, *Finding Your Seat at the Table: Roles for Librarians on Institutional Regulatory Boards and Committees* is one of few sources to find comprehensive information on librarians who serve IACUCs. Whether memorandum of understanding assisting with literature searches for animal alternatives or serving the IACUC in other capacities, librarians may receive a letter of thanks from their institution or simply have another mention of university service to add to their curriculum vitae. However, it can be said that there is also satisfaction felt among many librarians that they contribute to very important processes that help both animals and researchers, and ultimately, the advancement of medicine.

NOTES

1. Nuno Henrique Franco, "Animal Experiments in Biomedical Research: A Historical Perspective," *Animals* 3, no. 1 (March 2013): 252, https://doi.org/10.3390/ani3010238; "Nobel Prizes in Medicine," Foundation for Biomedical Research, accessed May 23, 2021, https://fbresearch.org/medical-advances/nobel-prizes/.

2. Foundation for Biomedical Research, "Nobel Prizes."

3. "Why Animal Research?" Stanford Medicine, accessed May 23, 2021, https://med.stanford.edu/animalresearch/why-animal-research.html.

4. Franco, "Animal Experiments," 239.

5. Franco, "Animal Experiments," 239–41.

6. Franco, "Animal Experiments," 252.

7. Steven J. Smith et al., "Use of Animals in Biomedical Research. Historical Role of the American Medical Association and the American Physician," *Archives of Internal Medicine* 148, no. 8 (1988): 1849, https://pubmed.ncbi.nlm.nih.gov/3041941/.

8. Smith et al., "Use of Animals in Biomedical Research," 1849–50.

9. Thomas L. Wolfle, "50 Years of the Institute for Laboratory Animal Research (ILAR): 1953–2003," *ILAR Journal* 44, no. 4 (2003): 325, https://doi.org/10.1093/ilar.44.4.324.

10. Nelson Garnett, "Introduction: By Ladies and Gentleman, for Ladies and Gentlemen," *ILAR Journal* 52, Supplement (2011): 420, https://olaw.nih.gov/seminar/docs/ILAR%20J%2052(S).pdf.

11. Wolfle, "50 Years," 325.

12. Jeremy R. Garrett, *The Ethics of Animal Research: Exploring the Controversy* (Cambridge: MIT Press, 2012), 2, https://ebookcentral.proquest.com/lib/duke/reader.action?docID=3339419.

13. Wolfle, "50 Years," 325.

14. Garnett, "Introduction," 419.

15. "AALAS Timeline," American Association for Laboratory Animal Science, accessed December 30, 2020, https://www.aalas.org/about-aalas/history/timeline.

16. "History," American Association for Laboratory Animal Science, accessed December 30, 2020, https://www.aalas.org/about-aalas/history.

17. W. M. S. Russell and R. L. Burch, *The Principles of Humane Experimental Technique* (London: Methuen, 1959), 64, https://catalog.hathitrust.org/Record/001498811?.

18. Russell and Burch, *The Principles,* xiv.

19. Wolfle, "50 Years," 324.

20. "College History," American College of Laboratory Animal Medicine, accessed December 30, 2020, https://www.aclam.org/about/college-history.

21. Daniel Engber, "Where's Pepper?" *Slate,* June 1, 2009, http://www.slate.com/articles/health_and_science/pepper/2009/06/wheres_pepper.html.

22. Garnett, "Introduction," 419.

23. Eileen Morgan, Kim Taylor, and Venita Thornton, "Writing a Good Assurance," Office of Laboratory Animal Welfare, accessed December 30, 2020, https://olaw.nih.gov/sites/default/files/110609_Assurance_slides.pdf.

24. "Why Conduct Literature Searches for Alternatives?" USDA National Agricultural Library, accessed December 30, 2020, https://www.nal.usda.gov/awic/why-conduct-literature-searches-alternatives.

25. Atha Louise Henley, Kathrine MacNeil, and Gretchen Stephens, "The Veterinary Medical Libraries Section of the Medical Library Association: 1972-1998," accessed December 30, 2020, https://www.mlanet.org/p/co/ly/gid=31&fid=1128&req=load.

26. Melissa Ratajeski, email message to author, January 4, 2021.

27. Ibid.; Judy Burnham, Melissa Ratajeski, and Kristin Hitchcock, "New Special Interest Groups (SIGs)," *MLA News*, March 2009, 22, https://www.mlanet.org/p/do/sd/sid=203.

28. Melissa Ratajeski, email message to author, January 4, 2021.

BIBLIOGRAPHY

American Association for Laboratory Animal Science. "AALAS Timeline." Accessed December 30, 2020. https://www.aalas.org/about-aalas/history/timeline.

American Association for Laboratory Animal Science. "History." Accessed December 30, 2020. https://www.aalas.org/about-aalas/history.

American College of Laboratory Animal Medicine. "College History." Accessed December 30, 2020. https://www.aclam.org/about/college-history.

Burnham, Judy, Melissa Ratajeski, and Kristin Hitchcock, "New Special Interest Groups (SIGs)." *MLA News*, March 2009. https://www.mlanet.org/p/do/sd/sid=203.

Engber, Daniel. "Where's Pepper?" *Slate,* June 1, 2009. http://www.slate.com/articles/health_and_science/pepper/2009/06/wheres_pepper.html.

Foundation for Biomedical Research. "Nobel Prizes in Medicine." Accessed May 23, 2021. https://fbresearch.org/medical-advances/nobel-prizes/.

Franco, Nuno Henrique. "Animal Experiments in Biomedical Research: A Historical Perspective." *Animals* (Basel) 3, no. 1 (March 2013): 238–73. https://doi.org/10.3390/ani3010238.

Garnett, Nelson. "Introduction: By Ladies and Gentleman, for Ladies and Gentlemen." *ILAR Journal* 52, Supplement (2011): 419–21. https://olaw.nih.gov/seminar/docs/ILAR%20J%2052(S).pdf.

Garrett, Jeremy R. *The Ethics of Animal Research: Exploring the Controversy.* Cambridge: MIT Press, 2012. https://ebookcentral.proquest.com/lib/duke/reader.action?docID=3339419.

Henley, Atha Louise, Kathrine MacNeil, and Gretchen Stephens. "The Veterinary Medical Libraries Section of the Medical Library Association: 1972-1998." Accessed December 30, 2020. https://www.mlanet.org/p/co/ly/gid=31&fid=1128&req=load.

Morgan, Eileen, Kim Taylor, and Venita Thornton. "Writing a Good Assurance." Accessed December 30, 2020. https://olaw.nih.gov/sites/default/files/110609_Assurance_slides.pdf.

Russell, W. M. S., and R. L. Burch. *The Principles of Humane Experimental Technique*. London: Methuen, 1959. https://catalog.hathitrust.org/Record/001498811?.

Smith, Steven J., R. Mark Evans, Micaela Sullivan-Fowler, and William R. Hendee. "Use of Animals in Biomedical Research. Historical Role of the American Medical Association and the American Physician." *Archives of Internal Medicine* 148, no. 8 (1988): 1849–53. https://pubmed.ncbi.nlm.nih.gov/3041941/.

Stanford Medicine. "Why Animal Research?" Accessed May 23, 2021. https://med.stanford.edu/animalresearch/why-animal-research.html.

USDA National Agricultural Library. "Why Conduct Literature Searches for Alternatives?" Accessed December 30, 2020. https://www.nal.usda.gov/awic/why-conduct-literature-searches-alternatives.

Wolfle, Thomas L. "50 Years of the Institute for Laboratory Animal Research (ILAR): 1953–2003." *ILAR Journal* 44, no. 4 (2003): 324-37. https://doi.org/10.1093/ilar.44.4.324.

Chapter Six

Librarians and the IACUC

Practical Approaches for Performing Alternatives Searches and Providing Support

Melissa A. Ratajeski and Rebekah S. Miller

U.S. federal law requires that institutions engaging in animal research must have an Institutional Animal Care and Use Committee (IACUC), which is tasked with managing and caring for all laboratory or experimental animals.[1] Researchers working with animals must submit protocols detailing their experiments, and those protocols are reviewed and approved (or rejected) by members of the committee. Depending on the animals and procedures used, researchers may need to perform in-depth literature searching as part of their protocol, demonstrating that alternatives to painful and/or distressful procedures were investigated. Due to this focus on searching the literature, librarians are sometimes contacted by their local IACUC for their expertise; other times librarians might approach their committee to offer services. However, library support of IACUCs can vary, and a full blown "service" might not develop immediately. If you are interested in beginning to support your local IACUC, the services you provide might depend on whether or not you are asked to become an IACUC member (voting or nonvoting) and if there is an expectation that you will perform protocol review. Whether librarians are interested in becoming (or invited to become) committee members most likely depends on discussion with the committee chair or other involved parties. After these discussions, you might have a better feel for what type of support would best suit your institution. Otherwise, one way to determine the appropriate level of support is to attend several full committee meetings so that you can begin to understand the issues both researchers submitting protocols and committee members reviewing protocols are facing.

One simple approach for support is creating a tailored LibGuide or webpage that provides information on library resources and services.[2] Some librarians have created LibGuides with in-depth information, including checklists or quizzes, to assist researchers with IACUC-mandated literature searches (see

Table 6.1.   Recommended LibGuides

Institution	Link	Strengths
Duke University	https://guides.mclibrary.duke.edu/animalalternatives	Suggested keywords
University of California, Davis	https://www.library.ucdavis.edu/guide/animal-alternatives-searching/	Suggested databases Suggested MeSH
University of Denver	https://libguides.du.edu/alternatives	Suggested keywords Sample searches using MeSH
University of North Carolina at Chapel Hill	https://guides.lib.unc.edu/animal-alternatives/	Learning module aimed at researchers to help them perform their own searches for research alternatives

table 6.1). At the University of Pittsburgh we have information on our library website about the library's IACUC services along with search tips and a contact form.[3] Additionally, the IACUC liaison librarian's contact information is included in the protocol system, so that researchers know where to get help at the point of need. One thing to keep in mind is that there are different subsets of people involved in animal research at each institution—including but not limited to primary investigators (PIs), administrators, graduate students, committee members, animal care technicians, and veterinarians—and each subset of patrons may have slightly different needs. To begin to get to know the committee members you might also participate in facility inspections, as Public Health Service (PHS) requires that all animal facilities are inspected at least once every six months.[4] These strategies will increase your profile, and hopefully your recognition as a resource as well.

While what you choose to do will be based on local needs and possibly library capacity, we suggest three areas for librarian IACUC involvement: completing searches, reviewing searches, and instruction.

LITERATURE SEARCHES

Librarians have a range of potential involvement with IACUC alternatives searches. (Note that we provide detail on what these searches are and what they entail later in this chapter.) Librarians might be asked to complete searches and deliver results, review PI-performed searches for investigators or the committee, or provide instruction on how to do searches. The librarian's level of involvement can depend on many factors and can differ on a research team–basis (i.e., the librarian does all searches for some teams but just provides guidance to others). For clarity we have split the types of searches into two—for the committee and for primary investigators.

Conducting Searches for the Committee

Literature searches requested by the committee will be much different than the alternatives searches required for protocols. Searches requested by the committee, in our experience, tend to be questions of policy or to support standard operating procedure updates and revisions. Additionally, the committee might ask you questions about best practices, established animal models, or animal welfare issues. Sometimes the committee might request an on-the-fly search during a full committee meeting, if there are questions during debate or discussion of a protocol, especially protocols that include animals in pain class E (in which anesthetics, analgesics, tranquilizers, or other pain/distress relieving measures are not used).[5] It can be helpful to have a working knowledge of available resources for alternatives searching—biomedical databases as well as more specialty resources. (See "Where to Search" section of this chapter for details on databases.)

Similarly, your institution's veterinarians or animal caretakers might ask you to conduct searches. These searches might be similar to those done for the committee itself, but can often be about environmental enrichment, single housing, or animal welfare.

Conducting Searches for Principal Investigators

PIs may ask librarians to complete the literature searches needed to satisfy the requirements of the Animal Welfare Act (AWA)[6] and USDA Policy 12[7] for them. When this request is made, the librarian must make explicit that the research group is still responsible for reviewing the literature retrieved and writing the necessary narratives. These searches are different from "regular" literature searches, and so this section will provide guidance on how to prepare for and conduct an alternatives search.

Your first step should be to make sure you understand exactly what the AWA requires, the definitions in USDA Policy 11,[8] and the guidance given in Policy 12.[9] These were already covered in a previous chapter, but we will review briefly here. The AWA requires that the principal investigator considers alternatives to any procedure likely to produce pain or distress in an experimental animal and provides a written narrative of the methods and sources used to determine that alternatives are not available. The AWA has a specific definition for an animal, limiting the scope to warm-blooded animals excluding rats (of the genus *Rattus*), mice (of the genus *Mus*), and birds bred for research.

USDA Policy 11 defines a painful procedure as "any procedure that would reasonably be expected to cause more than slight or momentary pain and/or distress in a human being to which that procedure is applied"; some examples given are terminal surgeries or skin irritation tests.[10] No definition of distress is provided in the policy but examples such as food or water regulation and immobility are provided in the guidance.

USDA Policy 12 provides minimal guidance on the requirement to provide a written narrative of the consideration of alternatives:

- "[T]he performance of a database search remains the most effective and efficient method for demonstrating compliance with the requirement." When a database is used, the narrative must include at minimum the name of databases used, search date, time period covered, and the terms and/or search strategy used.
- "[I]n some circumstances (as in highly specialized fields of study), conferences, colloquia, subject expert consultants, or other sources may provide relevant and up-to-date information regarding alternatives in lieu of, or in addition to, a database search."

Searching for alternatives means considering each of the 3Rs (refinement, replacement, reduction), as discussed by Russell and Burch in their definitive publication *The Principles of Humane Experimental Technique*.[11] As a reminder, definitions of these terms as they relate to alternatives are as follows:

1. Refinement: Employ techniques that reduce pain and distress
2. Replacement: Substitute animals with non-animal methods (i.e., computer simulation) or lower animal species
3. Reduction: Minimize the number of animals used, without jeopardizing statistical validity

Once you have a solid understanding of the federal mandates and guidance the next step is to understand how your own institution may enforce/interpret these, as protocol requirements vary by institution.[12] Requirements should be outlined in the protocol submission application or you may need to consult with the IACUC chair for details.

Here are some considerations:

- For which animal models does your IACUC require a literature search? In our experience some institutions may only require literature searches for animal models that are covered under the AWA, rather than all animals used in your research facility, excluding rats and mice for example. Other institutions may require searches for all animal use, regardless of species.

- How many databases are researchers required to use? Is one sufficient or are multiple required? Can documentation of conference attendance or expert consultation be used in lieu of, or in addition to, a database search as noted in USDA Policy 12?
- Are there restrictions put into place regarding which databases can/should be utilized? For example, is Google Scholar acceptable?
- Does the IACUC prescribe how many years a new protocol search should cover? Can publication date limits such as ten years be used?
- The guidance in Policy 12 simply states "keywords and/or the search strategy" should be documented. Does your institution require one over the other? University of Pittsburgh, for example, requires that a search strategy be uploaded with protocol submission. This allows reviewers to verify that keywords and subject headings are properly being combined and used.

Next, consider how the type of protocol submission the researcher is completing will impact your search. Within institutions, no animal experimentation or use (for example in a surgical training continuing education course) is permitted without written approval by the IACUC. This rule requires researchers to complete a protocol application for review, often in an online submission system. Only once a protocol has received the proper approval can the researcher conduct the experiments described within the protocol, using the number of animals and species justified.

Your institution likely has different types of IACUC protocol submissions as shown below. Note depending on which submission the researcher is completing, your search criteria may differ, or a search may not be required.

New application: Should include such details as the identification of the species, details of the research, justification for requested number of animals, description of the procedures, analgesics/anesthetics to be used, and euthanasia methods. New applications are approved for a time period of three years, as specified by the PHS Policy.[13] The search for this submission will be the most exhaustive both in terms of keywords and subject headings used as well as the time period covered.

3-Year Renewal: The PHS Policy requires de novo review of a continuing study protocol after three years. The same level of detail should be provided as in a new application but may include review of work already done to this point. If no substantial changes have been made to the protocol, a search supporting a three-year renewal can often "re-use" past searches with a few caveats. Librarians should always review to see if changes/updates in controlled vocabularies like MeSH (Medical Subject Headings) would impact the search, if there are new keywords used by the discipline to describe the area of research or a procedure used in the project, and/or if changes in the database require changes in search syntax.

A renewal search can be narrower in the time period covered. To narrow the search, librarians can rerun the original search (or the modified version) and either limit the publication date to the past three years or use fields such as PubMed's Entry Date (aka [EDAT] or Entrez Date)[14] to retrieve citations that have been added into the database since the last time the search was conducted. Note the latter may pull in older articles that were just added to the database more recently and would therefore be more comprehensive.

Modification Request: Both the AWA and PHS Policy require that the IACUC review and approve any proposal modifications. Some requests may be deemed "administrative requests" such as changes in personnel, funding source, or protocol name change. Other changes often directly relate to the research, including a procedural change that logically relates to a specific aim of the original protocol or a modification of analgesics/anesthetics. A modification request may just require a search specifically related to the modification. For example, just using terms for the new painful/distressful procedure or anesthesia drug proposed versus all the procedures/drugs already outlined in the experimental design of the original protocol.

Annual Review: Even though a protocol receives a three-year approval time period, the AWA requires ongoing reviews of the activities involving animals. These are to be at "appropriate intervals as determined by the IACUC, but not less than annually."[15] Consideration of alternatives are not required by IACUCs at the time of each annual review of the animal protocol, therefore searches are most likely not required by your institution for this submission type.

Finally, it is time to delve into the researcher's protocol and prepare for a reference interview. We have found that *before* meeting with a researcher for a reference interview the librarian should take time to review the protocol submission in its entirety. This approach allows you time to conduct some preliminary searches that may be needed to understand the research area, animal model, or the procedures and to formulate your specific questions.

We feel strongly that you should review the protocol yourself rather than relying solely on the researcher to provide you with a description of the area of research and experimental design. In our experience relying on a researcher's description can insert bias into the search, with researchers inadvertently leaving out details such as not including distressful procedures or those completed under anesthesia.

How to access this protocol varies on whether your institution uses an online submission system or a paper-based process.[16] If an online system, access may rely on what permissions the system allows and/or IACUC policies regarding access. At the University of Pittsburgh both of the librarians, who are nonvoting members of the IACUC, have read permission access to all of the protocols at the institution. This allows us to read the current protocol and/or past versions that include details on study aims, animal models, experimental design, anesthesia and analgesia, euthanasia, and past literature searches. We can also review any modifications that might have been submitted and see contact information for the principal investigator, surgical staff, and general staff. In the past these contacts proved valuable to get further clarification needed for searches.

When you begin your discussions regarding IACUC library support, ask what the protocol submission process looks like and insist that you have access if an online system. If your system is paper-based, you will most likely need to rely on the researchers sending you a copy of the protocol. These communications should be kept private and not forwarded to unauthorized individuals.

Once you have access to the protocol some suggestions include the following:

- Read the protocol in its entirety at least once. There are some sections that may seem irrelevant; however, it's our experience that researchers include 3R-related details throughout.
- As you are reading the protocol, make three lists with details on aims of the research, all procedures used, and anesthesia/analgesia used.
- Make notes on the specific animal model/s being used. Include details such as common name, species name, strains, age, and sex.
- After reading the protocol conduct some preliminary searches to help you determine what questions you might have for the researcher. Review controlled vocabulary and any citations given in the protocol.

The Reference Interview

It is now time to take all the information you have gathered and conduct a reference interview. You may find that an interview for a 3R-related search varies some from interviews for other reference questions. A "reference interview is the conversation that occurs between the librarian and user, in which the librarian asks questions aimed at determining the type and amount of information needed and how it is to be used."[17] Because the scope of the 3R search is already defined, determining the type of information and how it will be used is often not necessary, once you know the protocol submission type, as mentioned above.

First things first: make sure the person that you are conducting the reference interview with has the level of knowledge and expertise to answer the questions you have. An administrative assistant may have approached you about doing the search, but might not be positioned to effectively answer your questions. You should feel empowered to ask to speak with the contacts that you have identified while reading the protocol or others that might be able to provide the level of detail needed.

Most of your time in the interview will be used in clarifying terminology for the research aims and determining what procedures are painful and/or distressful. Although it is best practice to ask open-ended questions during an interview ("questions that cannot be answered by 'yes' or 'no'"),[18] for these searches it is appropriate to share the list of procedures you created during your review of the protocol and ask for verification/validation of the appropriate procedures to include in the refinement search section. You can then also ask for the context of "why the procedure is being performed . . . the expected outcome,"[19] and possible known alternatives. It can also be useful to ask how much literature is expected to be retrieved, especially for articles that relate to similar research aims. New areas of research and novel procedures will obviously have lower retrievals. In addition, there should be discussion of recall versus precision of the search. Management of expectations regarding the number of citations that will be retrieved versus the number the team is willing to review may be needed. It should also be made clear to the PI that it is their responsibility to review the citations, consider any alternatives uncovered, and write the narratives. Here are some other suggested questions to ask (even if you might know the answers from reading the protocol):

- Can you describe the research aims?
- What animal model/s are typically used for this research?
- Is the animal model/s that you are using novel for this area of research?
- Could an animal lower on the phylogenetic scale be used?
- Can you walk me through the experimental design?
- What terms would you search with for this topic/procedure?
- Do you have any relevant citations that you could share?
- What anesthesia/analgesia will the animals be given?
- How would you like to receive the results of the search?
- What is your timeframe for needing the results?

The Animal Welfare Information Center (AWIC) has an available worksheet that could be useful to adapt as appropriate for use during reference interviews.[20]

Caveats before Searching

Before discussing how to run these searches it is worth considering the challenges that are inherent in conducting them. Two major concerns are the lack of methods reporting in manuscripts and the lack of indexing of articles in databases to facilitate retrieval.

As Carbone and Austin state, "A database search is no stronger than the literature in the field it searches."[21] One of the major concerns in this area is the lack of transparent reporting of details in the literature that would actually allow a researcher to consider alternatives. For example, if an article does not mention anesthesia or analgesia should the reader assume it was not given, is not necessary, or even that its use could alter the results?[22] Also to add to the problem, even if details are reported they are often not done so in a searchable field such as an abstract,[23] so useful articles might not be easily findable.

In 2010, the ARRIVE (Animal Research: Reporting *In Vivo* Experiments) guidelines[24] were developed in the hopes of promoting consistent and comprehensive reporting. They consist of a checklist that outlines the minimum information that should be included within the abstract, methods, results, and discussion sections of a research article where animal models were used. These include animal number, animal specifics (such as species, strain, and sex), housing information, and description of all procedures carried out (including anesthesia/analgesia).[25] Adherence to these standards, however, has been low.[26] In 2020, an easier to follow and streamlined version of the guidelines was published with the hopes of encouraging researchers (and journals) to use and promote them.[27] The ARRIVE 2.0 guidelines highlight what is most important to report, breaking the original checklist into two groups— Essential 10 and Recommended Set, with experimental procedures included in the essential list.[28]

Supporting documentation[29] for the guidelines states that the "What, When, Where, and Why" of a study should be included in enough detail for replication. However, the Why section simply states "provide rationale for procedures,"[30] with no push or suggestion to include whether alternatives were considered. So although the use of the ARRIVE guidelines is a good first step towards research transparency, their use does not encourage researchers to detail whether a particular procedure was chosen because it would reduce pain and distress or was less invasive than other potential methods.[31] We believe examples showing that alternatives were considered should be provided in the ARRIVE "explanation and examples" document[32] to encourage researchers to report this level of detail and to normalize the importance of careful consideration of alternatives. Chilov et al.[33] take this idea one step further, recommending that articles be indexed with controlled vocabulary (and author provided keywords) when there is any discussion of alternatives in an article, thus making this information easier for searchers to find.[34]

Unfortunately, the lack of indexing across databases is one of the biggest issues with trying to retrieve articles for alternatives. This includes the "lack of 3 R related terms in the thesauri used to index them"[35] as well as restrictions on how many controlled vocabulary terms can be used per article.[36] Compounding the problem, guidance given to indexers states that they should not include terms that are not connected with the research aims, thus removing any indexing for methods or alternatives.[37] For example, the indexing policy related to techniques and equipment in the MEDLINE Indexing Online Training Course[38] states, "Techniques are described routinely in almost all articles. If a technique is merely mentioned and not discussed, it will be considered third tier and not indexed at all."

Therefore methods mentioned in the full text of the article often will not be used when indexing the manuscript. Thus it is important that researchers include details concerning methodology in searchable fields like the article abstract (note the new ARRIVE 2.0 guidelines only recommend including this level of detail in the abstract rather than labeling it essential)[39] or author supplied keywords, which databases such as PubMed display and index. Without this added information in searchable fields there is no easy way to retrieve the citation from a database when looking for alternatives, unless utilizing a database that allows full text searching. Databases that allow full text searching (e.g., PMC or Europe PMC) are fewer in number, however, and considerations on how to adjust search strategies to optimize precision versus recall are needed.

Defining Search Concepts and Conducting the Searches

Often when describing searching for alternatives the term *search* is used, insinuating one search is all that is required. Realistically, however, this search is broken up into many concepts and "mini-searches" including the following:

- Animal model
- Similar research
- Reduction
- Refinement (including anesthesia/analgesia)
- Replacement

In this section we will suggest approaches, keywords, and medical subject headings related to these concepts; however, they should not be considered prescriptive. In addition, we are assuming a basic level of search expertise with using Boolean operators and controlled vocabulary and understanding of truncation and phrase searching. For database suggestions see table 6.3 in the "Where to Search" section of this chapter. The focus here will be on searching MEDLINE. Reviewing related literature[40] and the selected LibGuides included in table 6.1 would also be valuable. The LibGuides include selected MeSH and keywords beyond what we discuss in this chapter.

It should be noted here that we do not treat these searches as systematic reviews, although there is obvious value in doing so[41]; rather we try to find an acceptable balance between precision and recall. If the researcher is overwhelmed by an excess number of citations or becomes frustrated by high recall and low precision, your work may be for naught and the research and animals could be negatively impacted. Therefore you may need to utilize methods such as limiting subject headings to the main topic of the article or using subheadings if the retrieval numbers are too high.

As you begin to build your searches, list the keywords (consider alternate spellings like British or variant word endings), synonyms, abbreviations, and available controlled vocabulary for each concept. While a controlled vocabulary is specific to the particular database that you are searching, reviewing the thesaurus for other databases could be useful in finding related terms or synonyms. The three most used in this area are the PubMed MeSH Database,[42] AGRICOLA Thesaurus for Animal Use Alternatives,[43] and CAB Thesaurus.[44]

As discussed under the caveat section, these searches can be challenging to conduct and have limitations. Often, depending on the resource and which concept you are building, using keywords with field tags may be the most successful approach. Field tags will vary between databases but for PubMed and Ovid Medline we have often used tiab (PubMed) and mp (Ovid). See below for examples from each database:

- "Title/Abstract [TIAB]: Searches in a citation's title, collection title, abstract, other abstract and keywords."[45]
  Example: enrichment[tiab]
- "Multi-purpose .mp: Searches in a citation's title, original title, abstract, subject heading word, keyword heading word, name of substance and a few others."[46]
  Example: enrichment.mp.

When using keywords we suggest you use a database that includes a proximity operator (unfortunately, PubMed does not have this option). Depending on the database this might be called an adjacency, near, or within operator.

They function by looking for words in relation to one another, separated by a certain number of words, as defined by the user. Depending on the operator, the exact word order may be required. Check each database's help section to verify both the correct syntax and functionality.

Most often at the University of Pittsburgh we use the Ovid platform to search Medline and AGRICOLA. This platform has the adjacency operator as defined below.[47]

- "Defined adjacency operator (ADJn): retrieves records that contain search terms within a specified number (n-1) of words from each other in any order (stop-words included). To use the adjacency operator, separate your search terms with ADJ and a number from 1 to 99 as explained below:

  ADJ1 Next to each other, in any order
  ADJ2 Next to each other, in any order, up to 1 word in between
  ADJ3 Next to each other, in any order, up to 2 words in between."

This operator can be combined with Boolean operators and truncation symbols as shown in the below examples, which illustrate various ways to search for the concept water regulation/reinforcement that is often included as a distressful procedure for non-human primates.

Examples: (water adj2 regulate).mp.
              (water adj2 regulat$).mp.
              (water adj2 (regulat$ or reinforc$)).mp.

Other search tips include these:

- Consider using subheading qualifiers[48] to increase precision of searches. Such related ones include the following:

Administration & Dosage
Adverse Effects
Education
Instrumentation

Methods
Prevention & Control
Surgery
Veterinary

- Subheadings are associated with particular MeSH terms; however, you can "float" subheadings as well. Floating retrieves citations where the "floated" subheading is attached to any MeSH term used to index the record. The syntax for this varies between databases so consult the help documentation.
- Use caution when considering the animal limiter.[49] Sometimes the term *animal* is not included on records and the article is indexed with the more specific animal term (e.g., *swine*). Also note that in Medline the animal limit is based on indexing and recent studies that have yet to receive indexing terms will be excluded (as will ones that will never be indexed).
- Some search platforms, like Ovid, allow users to search multiple bibliographic databases simultaneously. However, we advise against this practice, and suggest that each database be individually searched. Doing so will allow advanced search techniques to be used and each database's controlled vocabulary, fields, limits, and subheadings to be explored and used most effectively.
- As you build your searches keep in mind that the authors of biomedical literature are often writing solely to communicate their research results and will rarely use alternatives terms in an abstract. Thus, be cautious about ANDing broad alternatives terms (like *reduce*, *refine*, *pain*, or *alternative*), with procedures or areas of research as these might artificially limit your results.

Considerations for Each of the Search Concepts

As previously mentioned, alternatives searching is comprised of several different, smaller searches. We will discuss considerations for each of those searches now. At the University of Pittsburgh, we complete a search for similar research in addition to separate searches for each R concept. We remove any duplicate citations retrieved by multiple searches before sending to the researchers. (See the "Output/Delivery of the Searches" section of this chapter for details on how we do this.)

Animal Model Used in the Protocol

(See Statements 1–3 of table 6.2.)

The animal model search is often combined with the refinement and anesthesia/analgesia concepts but could also be used to narrow down results in the similar research search. Terms should include species, common and scientific name, and strain (as pertinent). When searching animal model keywords, keep in mind authors often use spelling variants, synonyms for the same species, or mention the species term in either the singular or plural.

Similar Research

(See Statements 4–8 of table 6.2.)

Librarians new to supporting IACUCs often find completing similar research searches the most comfortable, as they are reminiscent of searches one might conduct to support a narrative review on a field of study as a liaison or while on the reference desk. Depending on how the searches are structured and whether or not the animal model concept is included, the citations retrieved

- can provide a basic understanding of the literature published in a particular field and can help determine if there is unnecessary duplication;
- can be reviewed to see if any further reduction in animal numbers (n's) can occur by comparing the n's used in retrieved studies (reduction);
- detail techniques and anesthesia/analgesia used (refinement); and
- outline commonly used species or other methodology (replacement).

To construct the similar research search consider the objective of the study. Related terms should consider the discipline or subdiscipline of the area of study and the system or part of anatomy to be studied. Consider scientific and common names of any diseases, conditions, tissues, systems, or cell lines.

Reduction

(See Statements 21–24 of table 6.2.)

Although the objective of this search is to make sure that researchers reduce animal numbers as much as possible it is important that this is done without jeopardizing statistical validity. As noted above, the main way that a reduction search is conducted is by looking for articles within the research area using the same animal model as proposed in the protocol. The n's used in each study should then be reviewed and compared with the proposal to deter-mine the appropriate number of animals required.[50] If there are drastic differ-ences the researcher should reexamine or write a justification as appropriate.

Other approaches that we use to search the literature for citations relat-ing to reduction are combining the search hedges used in similar research to describe the research area with medical subject headings such as "research design," "animal use alternatives," and/or "sample size" or the AGRICOLA controlled vocabulary term "animal use reduction."

One other search strategy to consider is to complete some generic keyword searching in the title and heading word field to retrieve general articles about reduction.[51] Since this strategy does not include terminology specific to the protocol it could be used as a search hedge for every protocol.

Ovid Examples: ((animal$) adj3 (reduc$ or less$) adj3 number$).ti.

((animal$) and reduc$).ti. and (alternatives or experimentation).hw.

These are a sampling of article titles retrieved searching in this manner:

- When 3 Rs meet a forth R: Replacement, reduction and refinement of animals in research on reproduction. PMID: 30951977.
- The Sharing Experimental Animal Resources, Coordinating Holdings (SEARCH) Framework: Encouraging Reduction, Replacement, and Refinement in Animal Research. PMID: 28081116.
- Animal "models": How a mechanistic approach can reduce suffering and improve translatability. PMID: 28816054.
- Replacing, reducing and refining the use of animals in tuberculosis vaccine research. PMID: 27667476.

Refinement

(See Statements 9–20 of table 6.2.)

The refinement search is used to determine if any techniques that reduce pain and distress are possible and/or if current procedures could be refined accordingly. There is much to consider in this search and as Wood and Hart note, "the 3Rs are very open ended and any improvement at any level would be considered an 'alternative.'"[52]

Terms/related concepts to use in a search are as follows:

- names of specific experimental and/or surgical techniques or procedures
- environmental enrichment
- housing conditions
- husbandry or handling techniques (including operant conditioning/reinforcement training)
- anesthesia/analgesia (including classes and trade names of any drugs)
- generic and proper names of chemical compounds
- staff training

As noted above, a similar research search can help identify refinements for procedures in protocols. A PI may encounter a "more humane" or refined method they did not know about when reviewing the methods of studies with similar experimental design.[53] In addition these three approaches should be considered and possibly employed:

1. *List every procedure in the protocol that has the potential to cause pain or distress*. Identify possible keywords for each technique and determine if a subject heading for each procedure exists in the database that you are using. Complete a search for each of these and possibly limit to the animal model being used. Unless you are getting an unwieldy number of results do not combine with area of research. For example, lavages are done

in both tuberculosis and simian immunodeficiency virus research and a modification developed by one group could be relevant to the other.

2. *Conduct a search for known alternatives similar to the above.* For example, if a protocol includes routine blood draws under anesthesia searching for operant conditioning techniques for blood draws would also be useful.

3. *Search broadly for each technique.* This may include searching with broader MeSH terms to uncover alternative procedures that may not have been previously considered because the researcher was unaware of them. Chilov et al.[54] provide this illustration: "[I]f an animal protocol involves ear tagging as a method of animal identification, searching on the MeSH term 'Animal Identification Systems' combined with the MeSH term that describes the animal type (class, order, species, etc.) will return articles on ear-tagging techniques, as well as other options that differ from the ones that the researcher originally intended to use."

Replacement

(See Statements 25–31 of table 6.2.)

The replacement search looks at whether you can substitute the proposed model with non-animal methods (e.g., in vitro techniques, computer simulation) or with a lower animal model on the phylogenetic scale. The similar research search may uncover some of these replacement methods depending on the concepts you included (as long as you did not include the animal model).

Another approach would be to combine terms that describe the research aims with potential alternatives, making sure not to duplicate the same search as that already completed for the similar research search. Although the exact animal model used in the protocol should not be included in this portion of the search, broader MeSH such as animal models, animal disease models, or another species that is known to be a suitable animal model in the area of research could be included.

Choosing replacement terms that are relevant to the area of study is important. For example, for a protocol that uses animal models for a surgical training course the terms *in vitro techniques* or *cultured cells* are not appropriate—rather *cadaver*, *mannequin*, or *simulation* might find more reasonable replacements.

CASE STUDY

This case study will demonstrate some of the search approaches discussed above. It is simply a sample and should not be considered comprehensive for the topic. Ovid Medline was used to allow for adjacency searching.

Protocol topic: To develop safe and effective tuberculosis vaccines using a rhesus macaque animal model.

For:	See (in Table 6.2):
Animal model	Statements 1–3
Similar research	Statements 4–8
Refinement concept	Statements 9–20
Reduction concept	Statements 21–24
Replacement concept	Statements 25–31

Table 6.2. Ovid Medline Search Strategies

#	Searches
1	exp macaca/ or macaca fascicularis/ or macaca mulatta/
2	(monkey$ or non-human primate$ or nonhuman primate$).mp.
3	1 or 2
4	tuberculosis, pulmonary/ or mycobacterium tuberculosis/
5	tuberculosis/ or latent tuberculosis/
6	4 or 5
7	exp tuberculosis vaccines/ or bcg vaccine/
8	3 and 6 and 7
9	(gastric adj3 aspirat$).mp. or gastric lavage/
10	Lymph Node Excision/
11	lymph node biops$.mp.
12	Blood Specimen Collection/[Methods]
13	(blood adj2 (sampl$ or collect$ or draw$)).ti.
14	((pair$ or social or group$) adj2 hous$).mp.
15	environment design/
16	Self-Injurious Behavior/[Prevention & Control]
17	ketamine/ or ketaset.mp.
18	*anesthesia/
19	or/9-18
20	3 and 19
21	(animal$ and reduc$).ti. and (alternatives or experimentation).hw.
22	Sample Size/ or Research Design/
23	6 and 7 and 22
24	21 or 23
25	Disease Models, Animal/
26	models, animal/
27	exp In Vitro Techniques/
28	"animal use alternatives"/ or animal testing alternatives/
29	Computer Simulation/
30	or/25-29
31	6 and 7 and 30

A few things to note about this case example:

- Keywords were not used in the similar research search as the MeSH for this area of research is quite extensive.
- For the similar research search the animal model was included to reduce the number of results; because a separate replacement search is being conducted, this is appropriate.
- Subheadings were used as well as field tag searching.
- The subject heading anesthesia was limited to the main topic of the article aka "restricted to MeSH major topic" to narrow the number of results.
- Adjacency was used often in the refinement search where indexing for methods is limited.

WHERE TO SEARCH

Where you search will be dependent on your researchers' needs, the type of research they are doing, and your institutional subscriptions. As noted above, because the procedures needed for the refinement search are often not MeSH terms or other controlled vocabulary, it can be useful to use resources that allow adjacency for keyword searching or full-text searching. Your institution may also have restrictions on what can and cannot be searched. (Google Scholar, for example, is not allowed to be used for alternatives searching at our institution.) Note that you do not have to use the same resources across all 3Rs—for example you might use NORINA just for replacement.

See tables 6.3 and 6.4 for our suggestions of resources you may want to consider. This list is not comprehensive—there are many resources out there to assist with alternatives searching. We recommend familiarizing yourself with additional guides, including IACUC websites and LibGuides for more information. It is worth noting that there are also resources for specific animals that are not on our lists that may pertain to your researchers' animal models—for example, zfin.org for zebrafish.

Output/Delivery of the Searches (Data Management/ Documentation for Transparency)

It is important for researchers to know that they must keep documentation of IACUC searches, as these are eligible for federal review. Per Policy 12, researchers must record years and databases searched, date of search, and search strings and/or strategy.[56] Librarians completing searches for PIs should make sure they provide this information. (Of course local IACUC requirements may require that other information is documented as well.) Keep in mind that researchers may not be familiar with library databases or jargon, so it can be helpful to be very explicit when providing this information. For example, at the top of our Medline searches, we write,

> *A Medline search (using the Ovid interface), using the search terms listed below, was completed on [date] and limited to the years noted below.*

Librarians should also keep their own documentation, as researchers may return for an update for a three-year review or need a search for a new protocol that uses similar terms. Consider how you will save these searches—it may make sense to save them in database accounts, as applicable, as well as in documents on your computer. It is a good idea to use naming conventions and organization structures. For example, you might have a folder per PI, and a README file in each folder explicitly explaining your file structure or naming conventions. Consider also what you are documenting—in addition to search strings it may be useful to record why specific procedures or concepts did not end up in the final search, or why certain terms were not used. This will help you in the event of future searches or even just to answer any questions from researchers or the committee.

The method you use for delivering references will depend on how you normally deliver search results and PI preference. Some librarians create a bibliography document with abstracts included (often using citation manager software), some send citation manager files or libraries, and some send links directly to database results.[57] Prior to delivering results it may also be useful to set expectations—researchers may not realize just how many references they will receive. "The more realistic the users' expectations, the more satisfied they will be with the results."[58] At the University of Pittsburgh we deliver all of the retrieved citations to the PIs rather than hand-picking only a subset of articles for delivery.

Below we detail our approach for providing search results, but this is just one possibility out of many. We use the citation manager software Endnote heavily, but recommend utilizing the delivery method that works best for you. How we deliver search results using EndNote:[59]

1. Run the similar research search in the initial database searched and download the results into an EndNote library.
2. In EndNote, double check that the field "Name of Database" is properly labeled. If not, use the "Change Fields" batch edit feature to update the proper database name for each reference.

Table 6.3. Literature Databases

Name	Controlled Vocabulary	Platforms (not exhaustive, ones with URLS are freely available)	Adjacency Searching?	Notes
Agricola	NAL Thesaurus	NAL (https://agricola.nal.usda.gov), Ovid, EBSCO	Not in NAL; available on Ovid and EBSCO platforms	
APA PsycINFO	Thesaurus of Psychological Index Terms	APA PsycNet, EBSCO, Ovid, ProQuest	Yes	
Aquatic Sciences and Fisheries Abstracts (ASFA)	Two: AFSA Thesaurus; Taxonomic Terms	ProQuest	Yes	
BIOSIS Previews	BIOSIS Controlled Vocabulary	Web of Science, EBSCO, Ovid	Yes	Searching via natural language is recommended
CAB Abstracts	CAB Thesaurus	CABI, Ovid	Not in CABI; Yes in Ovid	
EMBASE	EMTREE	EMBASE.com, Ovid	Yes	Includes MEDLINE records
Europe PMC	Can toggle "synonym query expansion" (aka MeSH) on and off	Europe PMC https://europepmc.org	No	Full text searching; can limit search to specific parts of articles (i.e. methods). Includes Agricola, PMC, PubMed, preprint servers55, patents, and more

			Not in PubMed; available in Ovid and EBSCO platforms	
MEDLINE	MeSH	PubMed (https://pubmed.ncbi.nlm.nih.gov), Ovid, EBSCO	No	
Preprint Servers	None	See list here: https://asapbio.org/preprint-servers	No	Preprints are articles that have not yet been peer reviewed – they may have relevant information but should be marked accordingly
PubAg	NAL Thesaurus	PubAg https://pubag.nal.usda.gov	No	Database of articles from USDA researchers. No wildcards/truncation.
PubMed Central	MeSH	PMC https://www.ncbi.nlm.nih.gov/pmc	No	Full text searching; "Methods – Key Terms" is a searchable field
ScienceDirect	None	ScienceDirect	No	
Scopus	None	Scopus	Yes	Includes MEDLINE and EMBASE records
Web of Science Core Collection	None	Web of Science	Yes	
Zoological Record	Zoological Record Thesaurus	Web of Science, EBSCO, Ovid	Yes	

Table 6.4. Resources specific to alternatives searching

Name	About	Website	Notes
ALTEX (Alternatives to Animal Experimentation)	Journal	https://www.altex.org/index.php/altex	Open access; from CAAT Note that only searching one journal is not adequate for 3R searching
AltTox	Lots of information on alternatives or animal free toxicity research	http://alttox.org/	Collaboration between the Humane Society and Procter & Gamble. See also AFSA: https://www.afsacollaboration.org/
Animal Welfare Information Center (AWIC)	Training, legislation, search advice from NAL	https://www.nal.usda.gov/awic	
Animal Welfare Institute Refinement Database	"Refinement of Housing, Husbandry, Care, and Use of Animals in Research"	https://awionline.org/content/refinement-database	In addition to the database, has information on legislation, dissection alternatives, and much more
ATLA (Alternatives to Laboratory Animals)	Journal	https://journals.sagepub.com/home/atl	From FRAME Note that only searching one journal is not adequate for 3R searching
Canadian Council on Animal Care (CCAC)	Training modules	https://www.ccac.ca/en/training/modules/	

Organization	Description	URL	Notes
Center for Alternatives to Animal Testing (CAAT)	Programs, resources, videos related to alternatives	https://caat.jhsph.edu/	Based out of Johns Hopkins University; journal ALTEX
European Union Reference Laboratory for Alternatives to Animal Testing (EURL ECVAM)	Information on the 3 Rs, toxicity	https://ec.europa.eu/jrc/en/eurl/ecvam	
FRAME	Resources, training in the 3 Rs	https://frame.org.uk/	Publishes the journal ATLA
International Network for Humane Education (InterNICHE)	Includes resources, databases with alternatives and studies, more	http://www.interniche.org/	
National Centre for the Replacement, Refinement and Reduction of Animals in Research (NC3Rs)	3 Rs resources including webinars, guidance, and advice	https://www.nc3rs.org.uk/	
NORINA (A Norwegian Inventory of Alternatives)	Teaching materials database	https://norecopa.no/norina	
Procedures with Care	Tutorials and guides for handling animals, performing common procedures	https://researchanimaltraining.com/article-categories/procedures-with-care	

3. Use the "Change Fields" batch edit feature in EndNote to add a label to each reference relating to the search it was retrieved with—so, in this case, Similar Research.
4. Run the remaining three searches (refinement, reduction, and replacement), repeating steps 1–3. This results in a library in which each reference has one label: Similar Research, Refinement, and so on.
5. If using a second database repeat steps 1–4, importing the references into the same EndNote library as the first search.
6. Use the "Find Duplicates" feature to remove duplicate citations across the entire library. Note this will remove duplicates across labels. By default, EndNote will keep the first article added to a library, so if an article is brought in under both the refinement and reduction searches, the record you added second will be removed. If you want to keep all articles, skip this step.
7. Use the EndNote "Subject Bibliography" tool to create a bibliography. We created and use a customized IACUC output style that includes the abstract and the database the reference came from. The Subject Bibliography allows you to sort by label, so that the references are sorted into their categories—Similar Research or each of the 3Rs.
8. At the top of this citations document we note that overlap is to be expected and that researchers may find articles about reduction in the refinement section and vice versa.
9. Email the document with the citations along with a document for each database that details the search strategy used. Researchers can then upload the search strategies into the protocol system, and use the citations to both adjust their protocol and write their 3R narrative. We also send EndNote libraries or ris files on request.

One final consideration is whether the researchers you work with are proficient at locating full text. You might consider letting them know that you are able to help them locate PDFs of articles as necessary. It is also a good idea to provide information on what they should do with the results (review them and write the narrative), as well as that they should keep the search documentation in case of future review by the IACUC, USDA, or other parties. Developing boilerplate language you can send along with search results can preemptively answer questions researchers might have.

REVIEWING SEARCHES

Librarians are not always asked to conduct literature searches; sometimes they are asked to review them.

As a Reviewer for the IACUC

The IACUC may ask you to review protocols, which will require you to examine the entire research methodology, which we will not discuss. (Your IACUC should provide training.) When it comes to reviewing the search, however, there are obvious points to look for, including whether (and how) Boolean operators, controlled vocabulary, and natural language were used. If researchers are required to upload the search strategy with search result numbers look for inconsistencies or odd results. For example, is a common procedure listed as having very few results? Or, does a very specific, uncommon search word show millions of hits? These might be indications that the search is not being reported correctly. Look, too, at what concepts are being combined—are they logical? Are there concepts, drugs, or procedures missing? Use the "red flags" listed below in the instruction section to look for additional errors or omissions.

If the search is deemed inadequate, specific questions and/or comments should be provided, allowing the researchers to successfully edit and resubmit without the need for multiple rounds of questions that may delay protocol approval.

Reviewing for a PI

Prior to submission, a PI may ask you to review their searches. In order to provide appropriate feedback you will need to read their protocol or to have a very thorough understanding of their research and methods. That will enable you to suggest applicable controlled vocabulary and additional search terms. You can use the suggestions recommended in this chapter to offer corrections or modifications to the search. You may also want to explain how to combine concepts when searching. Other search tips researchers might not know about include truncation, explosions in PubMed/Medline, and limits.

INSTRUCTION

While performing or reviewing searches might make up the bulk of a librarian's support to the IACUC, we highly recommend providing instruction as well. There are various avenues for doing this.

Committee Instruction

One important type of instruction to consider is teaching the IACUC committee members to review 3R searches and look for errors. During this instruction the librarian's role on the committee (assuming membership) should be made explicit. There may be a false belief that the librarian will review all searches.

Committee members reviewing protocols may not be familiar with search syntax or controlled vocabulary, so it can be very helpful for you to provide them with red flags to look for in searches. These red flags can include basic search errors, like incorrect usage of Boolean operators and truncation, or categorizing Ovid Medline and PubMed as two different databases.[60] Other potential errors reviewers should look for include the following:

- Terminology from the wrong protocol (copied and pasted from a previous search)
- Not looking for distressful procedures (only painful)
- Zero search results (or one hundred million)
- Same search used for all 3Rs
- No drug trade/generic names used
- No synonyms used
- The word alternative(s) used instead of specific procedures
- "Terms [are] either too specific or too general" to be of use[61]

If errors are found or the searches are deemed inadequate reviewers should provide specific questions/comments to guide the researchers on the necessary changes. It should be clear to committee members how to refer a researcher to the library and if there is a mechanism for the librarian to be added onto the protocol as a reviewer if need be. Since IACUC members rotate, it can be useful to present this information on a semi-regular basis.[62] Your IACUC might also have additional ideas for instruction—perhaps general searching tips, in-depth database usage, or even citation management.

PI/Lab Instruction

An equally important group for librarians to provide instruction to is the researchers submitting protocols. You could offer a consulting service and work with one lab or PI at a time. A second option is to have seminars for groups of researchers—perhaps investigators who are new to the institution. You could partner with your IACUC—they probably do training already, get on their schedule—or you could partner with another entity at your institution. At the University of Pittsburgh, the Clinical and Translational Science Institute offers Responsible Conduct of Research trainings, and the IACUC librarians occasionally offer workshops through there. Not only does this help the researchers become better searchers, but it can provide you with an "in." Some of the instruction can be basic best practices while searching, while other aspects can be similar to the instruction provided to the committee—mistakes researchers should avoid, how to use controlled vocabulary, and even which databases to select. This is also a good time to talk about using the ARRIVE guidelines and data management, especially if your library provides data support. Additionally, we would recommend talking to this group about why searching is so important—it is not a regulatory hoop, but rather an opportunity to learn new techniques and methods, which can be difficult to keep up with. While alternatives searches are often treated as an afterthought, if they are performed during research planning stages, animal welfare can be considered up front, improving the research and strengthening grant applications.

MOVING FORWARD

As we have illustrated throughout this chapter there are many ways for librarians to become involved with their local IACUC—ranging in support levels and time commitments. While in some ways providing IACUC support resembles liaison work in other areas, understanding regulations, requirements, and issues surrounding these alternative searches are key to being successful. Although many librarians may initially feel outside of their comfort zone as they begin working in this area our hope is that this chapter's practical advice and approaches will allow librarians to get started and over time build confidence.

NOTES

1. USDA National Agricultural Library, Animal Welfare Information Center (AWIC), "Animal Welfare Act," https://www.nal.usda.gov/awic/animal-welfare-act

2. Virginia A. Lingle, "The Health Sciences Library and the IACUC in Animal Research: Collaboration for More Effective Use of Electronic Resources," *Journal of Electronic Resources in Medical Libraries* 5, no. 3 (2008): 243–259.

3. Health Sciences Library System, University of Pittsburgh, "IACUC Support," https://www.hsls.pitt.edu/iacuc-support.

4. National Institutes of Health (NIH), Office of Laboratory Animal Welfare (OLAW), U.S. Department of Health and Human Services (DHHS), *Public Health Service Policy on Humane Care and Use of Laboratory Animals.* NIH Publication No. 15-8013 (2015).

5. National Research Council (NRC) Committee on Regulatory Issues in Animal Care and Use, *Definition of Pain and Distress and Reporting Requirements for Laboratory Animals: Proceedings of the Workshop Held June 22, 2000*, Washington, DC: National Academies Press.

6. USDA National Agricultural Library, Animal Welfare Information Center (AWIC), "Animal Welfare Act."

7. National Research Council (NRC) Committee on Regulatory Issues in Animal Care and Use, *Definition of Pain and Distress and Reporting Requirements for Laboratory Animals.*

8. Ibid.

9. Ibid.

10. Ibid.

11. William Moy Stratton Russell and Rex Leonard Burch, *The Principles of Humane Experimental Technique*, Universities Federation for Animal Welfare (UFAW, 1992).

12. Susan C. Steelman and Sheila L. Thomas, "Academic Health Sciences Librarians' Contributions to Institutional Animal Care and Use Committees," *Journal of the Medical Library Association* 102, no. 3 (2014): 215–219.

13. National Institutes of Health (NIH), Office of Laboratory Animal Welfare (OLAW), U.S. Department of Health and Human Services (DHHS). *Public Health Service Policy.*

14. National Library of Medicine, "PubMed User Guide. Entry Date [EDAT]," PubMed, published 2020, https://pubmed.ncbi.nlm.nih.gov/help/#edat

15. USDA National Agricultural Library, Animal Welfare Information Center (AWIC), "Animal Welfare Act."

16. Swapna Mohan and Patricia L. Foley, "Everything You Need to Know About Satisfying IACUC Protocol Requirements," *ILAR Journal* 60, no. 1 (2019): 50–57.

17. Marie T. Ascher, "Reference and Information Services in Health Sciences Libraries," in *Health Sciences Librarianship*, ed. M. Sandra Wood (Lanham, MD: Rowman & Littlefield Publishers, 2014), 171–195.

18. Bryna Coonin and Cynthia R. Levine, "Reference Interviews: Getting Things Right," *The Reference Librarian* 54, no. 1 (2013): 73–77.

19. Adrian J. Smith and Tim Allen, "The Use of Databases, Information Centres and Guidelines When Planning Research that May Involve Animals," *Animal Welfare* 14(2005): 347–359.

20. Animal Welfare Information Center (AWIC), "Worksheet: Alternatives Literature Searching," published 2015, https://www.nal.usda.gov/sites/default/files/altwksht_jan2020.pdf.

21. Larry Carbone and Jamie Austin, "Pain and Laboratory Animals: Publication Practices for Better Data Reproducibility and Better Animal Welfare," *PLoS One* 11, no. 5 (2016): e0155001.

22. Ibid.

23. Emily S. Mazure, Melissa A. Ratajeski, Karen H. Gau, Erica R. Brody, Brian Di Pace, and Brandi Tuttle, "Searching for Animal Research Methodology: The Potential Impact of Differences Between the PubMed Record and Full-text Methods Section" (presented at the Medical Library Association Annual Conference, Seattle, WA, 2017).

24. Carol Kilkenny, William J. Browne, Innes C. Cuthill, Michael Emerson, and Douglas G. Altman, "Improving Bioscience Research Reporting: The ARRIVE Guidelines for Reporting Animal Research," *PLoS Biology* 8, no. 6 (2010): e1000412.

25. Ibid.

26. David Baker, Katie Lidster, Ana Sottomayor, and Sandra Amor, "Two Years Later: Journals are Not Yet Enforcing the ARRIVE Guidelines on Reporting Standards for Pre-Clinical Animal Studies," *PLoS Biology* 12, no. 1 (2014): e1001756; Julián Ernesto Nicolás Gulin, Daniela Marisa Rocco, and Facundo García-Bournissen, "Quality of Reporting and Adherence to ARRIVE Guidelines in Animal Studies for Chagas Disease Preclinical Drug Research: A Systematic Review," *PLoS Neglected Tropical Diseases*, 9, no. 11 (2015): e0004194; Nathalie Percie du Sert, Viki Hurst, Amrita Ahluwalia, Sabina Alam, Marc T. Avey, Monya Baker, William J. Browne, et al., "The ARRIVE Guidelines 2.0: Updated Guidelines for Reporting Animal Research," *BMC Veterinary Research* 16, no. 1 (2020): 242.

27. Percie du Sert et al., "The ARRIVE Guidelines 2.0."

28. Ibid.

29. National Centre for the Replacement Refinement and Reduction of Animals in Research, "ARRIVE Essential 10," published 2020, https://www.arriveguidelines.org.

30. Ibid.

31. Marina Chilov, Konstantina Matsoukas, Nighat Ispahany, Tracy Y. Allen, and Joyce W. Lustbader, "Using MeSH to Search for Alternatives to the Use of Animals in Research," *Medical Reference Services Quarterly* 26, no. 3 (2007): 55–74.

32. National Centre for the Replacement Refinement and Reduction of Animals in Research, "ARRIVE. Essential 10."

33. Chilov et al., "Using MeSH to search."

34. Ibid.

35. Krys Bottrill, "Information: Needs for the Future," *Alternatives for Laboratory Animals* 30, Suppl. 2 (2002): 145–149.

36. Ibid.

37. Ibid.

38. National Library of Medicine, "MEDLINE Indexing Online Training Course, Category E—Techniques and Equipment, Indexing Principles," published 2015, https://www.nlm.nih.gov/bsd/indexing/training/CATE_010.html.

39. Percie du Sert et al., "The ARRIVE Guidelines 2.0."

40. Chilov et al., "Using MeSH to search"; Carlijn R. Hooijmans, Alice Tillema, Marlies Leenaars, and Merel Ritskes-Hoitinga, "Enhancing Search Efficiency by Means of a Search Filter for Finding All Studies on Animal Experimentation in PubMed," *Laboratory Animals* 44, no. 3 (2010): 170-175; Annett J. Roi and Barbara Grune, The EURL ECVAM Search Guide Data Retrieval Procedures Basic Principles (Luxembourg, 2013)"; Daureen Nesdill and Kristina M. Adams, "Literature Search Strategies to Comply with Institutional Animal Care and Use Committee Review Requirements," *Journal of Veterinary Medical Education* 38, no. 2 (2011): 150-156.

41. Merel Ritskes-Hoitinga and Judith van Luijk, "How Can Systematic Reviews Teach Us More About the Implementation of the 3Rs and Animal Welfare?" *Animals* 9, no. 12 (2019): 1163; Marc T. Avey, Nicole Fenwick, and Gilly Griffin, "The Use of Systematic Reviews and Reporting Guidelines to Advance the Implementation of the 3Rs," *Journal of the American Association for Laboratory Animal Science* 54, no. 2 (2015): 153–162.

42. National Library of Medicine, "The MeSH Database," published 2020, https://www.nlm.nih.gov/bsd/disted/meshtutorial/themeshdatabase/index.html.

43. USDA National Agricultural Library, "NAL Agricultural Thesaurus and Glossary," published 2006, https://agclass.nal.usda.gov.

44. CABI International, "CAB Thesaurus," published 2019, https://www.cabi.org/cabthesaurus/mtwdk.exe?yi=home.

45. National Library of Medicine (NLM), "PubMed User Guide. Title/Abstract [TIAB]," PubMed, published 2020, https://pubmed.ncbi.nlm.nih.gov/help/#tiab.

46. Wolters Kluwer Health. "MEDLINE® 2020 Database Guide," published 2020, https://ospguides.ovid.com/OSPguides/medline.htm.

47. Ibid.

48. National Library of Medicine, "MeSH Qualifiers with Scope Notes," published 2020, https://www.nlm.nih.gov/mesh/qualifiers_scopenotes.html.

49. Hooijmans et al., "Enhancing Search Efficiency."

50. Smith and Allen, "The Use of Databases."

51. Melissa A. Ratajeski, "Reduction Search Strategies for Animal Research" (presented at the Medical Library Association Annual Conference, Philadelphia, PA, 2007).

52. Mary. W. Wood and Lynette A. Hart, "Searching for the 3Rs: Facilitating Compliance in the Bibliographic Search for Alternatives," *INSPEL* 35, no. 3 (2001): 191–198.

53. Chilov et al., "Using MeSH to Search."

54. Ibid.

55. Europe PMC, "Preprints in Europe PMC," published 2020 https://europepmc.org/Preprints

56. National Research Council (NRC) Committee on Regulatory Issues in Animal Care and Use, *Definition of Pain and Distress and Reporting Requirements for Laboratory Animals*.

57. Lee A. Vucovich, "Research Services and Database Searching in Health Sciences Libraries," in *Health Sciences Librarianship*, edited by M. Sandra Wood (Lanham, MD: Rowman & Littlefield Publishers, 2014), 196–225.

58. Stuart J. Kolner, "Improving the MEDLARS Search Interview: A Checklist Approach," *Bulletin of the Medical Library Association* 69, no. 1 (1981): 23–33.

59. Melissa A. Ratajeski, "Customizing EndNote for Institutional Animal Care and Use Committee (IACUC) Protocol Searches" (presented at the Medical Library Association Annual Conference, Chicago, IL, 2008).

60. Lingle, "The Health Sciences Library."

61. Ibid.

62. Ibid.

BIBLIOGRAPHY

Animal Welfare Information Center (AWIC). "Worksheet: Alternatives Literature Searching." Published 2015. https://www.nal.usda.gov/sites/default/files/altwk-sht_jan2020.pdf.

Ascher, Marie T. "Reference and Information Services in Health Sciences Libraries." In *Health Sciences Librarianship,* edited by M. Sandra Wood, 171–195. Lanham, MD: Rowman & Littlefield Publishers, 2014.

Avey, Marc T., Nicole Fenwick, and Gilly Griffin. "The Use of Systematic Reviews and Reporting Guidelines to Advance the Implementation of the 3Rs." *Journal of the American Association for Laboratory Animal Science* 54, no. 2 (2015): 153–162.

Baker, David, Katie Lidster, Ana Sottomayor, and Sandra Amor. "Two Years Later: Journals are Not Yet Enforcing the ARRIVE Guidelines on Reporting Standards for Pre-Clinical Animal Studies." *PLoS Biology* 12, no. 1 (2014): e1001756.

Bottrill, Krys. "Information: Needs for the Future." *Alternatives for Laboratory Animals* 30, Suppl. 2 (2002): 145–149.

CABI International. "CAB Thesaurus." Published 2019. https://www.cabi.org/cabthesaurus/mtwdk.exe?yi=home

Carbone, Larry, and Jamie Austin. "Pain and Laboratory Animals: Publication Practices for Better Data Reproducibility and Better Animal Welfare." *PLoS One* 11, no. 5 (2016): e0155001.

Chilov, Marina, Konstantina Matsoukas, Nighat Ispahany, Tracy Y. Allen, and Joyce W. Lustbader. "Using MeSH to Search for Alternatives to the Use of Animals in Research." *Medical Reference Services Quarterly* 26, no. 3 (2007): 55–74.

Coonin, Bryna, and Cynthia R. Levine. "Reference Interviews: Getting Things Right." *The Reference Librarian* 54, no. 1 (2013): 73–77.

Europe PMC. "Preprints in Europe PMC." Published 2020. https://europepmc.org/Preprints

Gulin, Julián Ernesto Nicolás, Daniela Marisa Rocco, and Facundo García-Bournissen. "Quality of Reporting and Adherence to ARRIVE Guidelines in Animal Studies for Chagas Disease Preclinical Drug Research: A Systematic Review." *PLoS Neglected Tropical Diseases* 9, no. 11 (2015): e0004194.

Health Sciences Library System, University of Pittsburgh. "IACUC Support." https://www.hsls.pitt.edu/iacuc-support.

Hooijmans, Carlijn R., Alice Tillema, Marlies Leenaars, and Merel Ritskes-Hoitinga. "Enhancing Search Efficiency by Means of a Search Filter for Finding All Studies on Animal Experimentation in PubMed." *Laboratory Animals* 44, no. 3 (2010): 170–175.

Kilkenny, Carol, William J. Browne, Innes C. Cuthill, Michael Emerson, and Douglas G. Altman. "Improving Bioscience Research Reporting: The ARRIVE Guidelines for Reporting Animal Research." *PLoS Biology* 8, no. 6 (2010): e1000412.

Kolner, Stuart J. "Improving the MEDLARS Search Interview: A Checklist Approach." *Bulletin of the Medical Library Association* 69, no. 1 (1981): 23–33.

Lingle, Virginia A. "The Health Sciences Library and the IACUC in Animal Research: Collaboration for More Effective Use of Electronic Resources." *Journal of Electronic Resources in Medical Libraries* 5, no. 3 (2008): 243–259

Mazure, Emily S., Melissa A. Ratajeski, Karen H. Gau, Erica R. Brody, Brian Di Pace, and Brandi Tuttle. "Searching for Animal Research Methodology: The Potential Impact of Differences Between the PubMed Record and Full-text Methods Section." Poster presented at the Medical Library Association Annual Conference, Seattle, WA, 2017.

Mohan, Swapna, and Patricia L. Foley. "Everything You Need to Know About Satisfying IACUC Protocol Requirements." *ILAR Journal* 60, no. 1 (2019): 50–57.

National Centre for the Replacement Refinement & Reduction of Animals in Research. "ARRIVE Essential 10." Published 2020. https://www.arriveguidelines.org.

National Institutes of Health (NIH), Office of Laboratory Animal Welfare (OLAW), U.S. Department of Health and Human Services (DHHS). *Public Health Service Policy on Humane Care and Use of Laboratory Animals.* NIH Publication No. 15-8013 (2015).

National Library of Medicine. "The MeSH Database." Published 2020. https://www.nlm.nih.gov/bsd/disted/meshtutorial/themeshdatabase/index.html.

National Library of Medicine. "MEDLINE Indexing Online Training Course, Category E - Techniques and Equipment, Indexing Principles." Published 2015. https://www.nlm.nih.gov/bsd/indexing/training/CATE_010.html.

National Library of Medicine. "MeSH Qualifiers with Scope Notes." Published 2020. https://www.nlm.nih.gov/mesh/qualifiers_scopenotes.html.

National Library of Medicine, "PubMed User Guide. Entry Date [EDAT]." Published 2020. https://pubmed.ncbi.nlm.nih.gov/help/#edat.

National Library of Medicine. "PubMed User Guide. Title/Abstract [TIAB]." Published 2020. https://pubmed.ncbi.nlm.nih.gov/help/#tiab.

National Research Council (NRC) Committee on Regulatory Issues in Animal Care and Use. *Definition of Pain and Distress and Reporting Requirements for Laboratory Animals: Proceedings of the Workshop Held June 22, 2000.* Washington, DC: National Academies Press.

Nesdill, Daureen, and Kristina M. Adams. "Literature Search Strategies to Comply with Institutional Animal Care and Use Committee Review Requirements." *Journal of Veterinary Medical Education* 38, no. 2 (2011): 150–156.

Percie du Sert, Nathalie, Viki Hurst, Amrita Ahluwalia, Sabina Alam, Marc T. Avey, Monya Baker, William J. Browne et al. "The ARRIVE guidelines 2.0: Updated guidelines for reporting animal research." *BMC Veterinary Research* 16, no. 1 (2020): 242.

Ratajeski, Melissa A. "Customizing EndNote for Institutional Animal Care and Use Committee (IACUC) Protocol Searches." Presented at the Medical Library Association Annual Conference, Chicago, IL, 2008.

Ratajeski, Melissa A. "Reduction Search Strategies for Animal Research." Presented at the Medical Library Association Annual Conference, Philadelphia, PA, 2007.

Ritskes-Hoitinga, Merel, and Judith van Luijk. "How Can Systematic Reviews Teach Us More About the Implementation of the 3Rs and Animal Welfare?" *Animals* 9, no. 12 (2019): 1163.

Roi, Annett J., and Barbara Grune. *The EURL ECVAM Search Guide Data Retrieval Procedures Basic Principles.* Luxembourg, 2013.

Russell, William Moy Stratton, and Rex Leonard Burch. *The Principles of Humane Experimental Technique.* Universities Federation for Animal Welfare (UFAW; 1992).

Smith, Adrian J., and Tim Allen. "The Use of Databases, Information Centres and Guidelines When Planning Research that May Involve Animals." *Animal Welfare* 14 (2005): 347–359.

Steelman, Susan C., and Sheila L. Thomas. "Academic Health Sciences Librarians' Contributions to Institutional Animal Care and Use Committees." *Journal of the Medical Library Association* 102, no. 3 (2014): 215–219.

USDA National Agricultural Library. "NAL Agricultural Thesaurus and Glossary." Published 2006. https://agclass.nal.usda.gov.

USDA National Agricultural Library, Animal Welfare Information Center (AWIC). "Animal Welfare Act." https://www.nal.usda.gov/awic/animal-welfare-act.

Vucovich, Lee A. "Research Services and Database Searching in Health Sciences Libraries." In *Health Sciences Librarianship,* edited by M. Sandra Wood, 196–225. Lanham, MD: Rowman & Littlefield Publishers, 2014.

Wolters Kluwer Health. "MEDLINE® 2020 Database Guide." Published 2020. https://ospguides.ovid.com/OSPguides/medline.htm.

Wood, Mary W., and Lynette A. Hart. "Searching for the 3Rs: Facilitating Compliance in the Bibliographic Search for Alternatives." *INSPEL* 35, no. 3 (2001): 191–198.

Part III

CREATING AND EXPANDING OPPORTUNITIES FOR LIBRARIES AND LIBRARIANS IN INSTITUTIONAL RESEARCH

Chapter Seven

Leveraging Participation into Partnership

How Serving on the IACUC Strengthens Librarian Relationships with Researchers

Andrea C. Kepsel

Serving on the Institutional Animal Care and Use Committee (IACUC) presents librarians with many unique opportunities for one-on-one and in-depth engagement with the research community. Some of these audiences may not be encountered in typical liaison or subject librarian activities. This chapter will examine the responsibilities a librarian may have as a committee member and how performing these duties can strengthen their performance as a librarian serving the various groups. The benefits of these encounters include stronger relationships that support better animal care and better research practices. While this chapter addresses some information already discussed in chapter 4 of this book ("The Institutional Animal Care and Use Committee: Membership, Responsibilities, and Roles for Librarians"), it does so to ground the reader and build subject matter fluency. This allows the conversation to move from the foundational to the actionable and aspirational.

THE ROLE OF THE LIAISON LIBRARIAN

Supporting an IACUC falls naturally within the responsibility of liaison or subject librarians. While the specific title and responsibilities may vary, the roles of these librarians increasingly incorporate research support[1] and have a goal of connecting users with information.[2] Building relationships is the essence of being a liaison librarian.[3] For the purposes of this chapter the title "liaison librarian" will be used as the focus is on establishing partnerships with IACUC members and researchers.

Typically, librarians on the IACUC are liaisons to biomedical science or basic science programs. Through their liaison work they may already be knowledgeable about their institution's research priorities and the resources necessary to support their IACUC duties. This does not preclude librarians in other subject areas or roles from becoming involved with the IACUC, but the benefits may be limited. While an in-depth science background is not absolutely necessary, some familiarity with the laboratory research process and commonly used sources of information is helpful. This knowledge will help the librarian be a productive member of the IACUC and proactively seek out opportunities for further collaboration.

RESPONSIBILITIES OF A LIBRARIAN ON THE IACUC

The role of a librarian member of an IACUC will be unique to each institution and responsibilities will vary.[4] Some librarians may be considered scientific members of the committee due to prior knowledge and experience, while others are classified as nonscientific or community members. Some may have full voting rights on the committee, and others may serve more of an advisory role. While this section will go into detail about various activities members of an IACUC may participate in, not all will be applicable in every setting, but perhaps they will provide inspiration to those looking to become more involved.

Literature Searches

The literature search is likely the most common role a librarian may have in relation to the IACUC, whether as a member of the committee or not. The Animal Welfare Act requires that researchers consider alternatives to any procedure likely to produce pain or distress, and that a written narrative description of the methods and sources is used to determine that alternative were not available.[5] A literature search is considered the best way to demonstrate this. Requirements for a literature search and documentation vary by institution, but in all cases, it is emphasized that the research literature should be consulted for alternatives. Librarians can use this to their advantage in establishing successful partnerships with researchers and IACUC committee members.

The literature search requirement for animal use protocols can be an easy way for librarians to start forging relationships with researchers at an institution. As experts trained in searching and locating information, librarians have an advantage over researchers in conducting a thorough search for alternatives. The immediate result is reducing the time that researchers spend learning what resources to use and how to conduct a proper search for alternatives. Over time this translates to cost savings and increased consideration of the 3Rs (replacement, reduction, and refinement).

Engaging with the literature search requirement for animal research presents numerous partnership opportunities for librarians. At least one librarian should be designated to provide assistance with literature searches for animal protocols, and their contact information should be included on any relevant animal research forms or websites so that it is easy for researchers to request assistance. If not already a member of the IACUC, the librarian should reach out to the committee and establish this line of communication. Referrals from the IACUC are a strong motivator for researchers to contact the librarian, and networking with the committee is an effective way to promote library resources and services to researchers.[6] It may be especially beneficial to encourage new researchers to contact the librarian when submitting their first animal protocol, which helps establish a relationship between the researcher and the library early on.

In recent years, there has been increased determination to provide support (driven by researcher requirements) and to pursue a more proactive model of engagement.[7] Assistance with the literature search requirement is one example of how librarians can respond to researcher needs and should be included in marketing and outreach efforts to researchers as an example of support that can be provided. In the process of performing searches for alternatives, librarians have an opportunity to learn about a researcher's work and scholarly interests. This can open up a conversation about additional support offered by the librarian, or resources available through the library, which may help in other areas of their work such as teaching or patient care. The librarian may set up search alerts and offer to send researchers new literature on alternatives as it becomes available, rather than waiting until a protocol comes up for renewal. Not only does this save time at renewal, it may assist researchers in applying new alternatives through the life of a protocol, helping their research remain current and increasing incorporation of the 3Rs. It may only take one thing to lead to a continual and mutually beneficial partnership between librarian and researchers, and fulfilling the literature search requirement is one place to start.

Protocol Review

One of the primary activities of an IACUC member is to review animal protocols and assure compliance to appropriate animal health and welfare guidelines. This includes reviewing of newly submitted protocols as well as those eligible for renewal. As an IACUC member a librarian may participate in reviewing of all animal protocols, only those that include a literature search for alternatives, or may just review the portions that include the searches themselves. The review of protocols can initially seem daunting, especially to those without a science background, but even nonscientific members can provide feedback on whether the protocol is transparent and understandable and whether or not alternatives have been considered.[8] The librarian provides a valuable perspective when reviewing a protocol for these factors, ultimately increasing the rigor and reproducibility of animal research being performed at an institution.

When reviewing protocols, the librarian will want to pay particular attention to any included literature searches and citations used for justification of the proposed protocol. A detailed and current search of appropriate sources is one of the best ways to assure that a researcher has considered current research on alternatives. A researcher should report what databases they searched, the keywords used, and when the search was performed. Review of the searches performed and literature used creates an opportunity for the librarian to communicate with the IACUC about recent research in alternatives and expectations for good searches. It also provides a path for the librarian to reach out to researchers and offer assistance when the committee determines the literature search is not adequate. Connections established at this point of need can lead to further communication between librarian and researcher, or librarian and committee, for additional support.

Meeting with Researchers

Meeting with researchers as part of an IACUC can expose the librarian to their research and make connections that may otherwise be missed. Meetings may take many forms, including new researchers attending an IACUC meeting to introduce their research to the committee, designated individuals meeting with those submitting protocols for the first time or starting new projects, or meeting with a researcher who may be experiencing issues or for whom the committee has questions to address regarding their protocol. By participating in these meetings, the librarian has an opportunity for in-depth conversations with researchers to further learn about their research and consider services and resources to support it. Involvement of the librarian in these activities also demonstrates that the librarian is a valued member of the IACUC and that researchers should reach out when questions arise.

Laboratory and Facility Inspections

Regular inspections by the IACUC of laboratory and other facilities where animal research takes place are required. Ability to participate in inspections varies by institution, but if the librarian is able to take part, it is a highly valuable experience. Inspections provide an opportunity to see, firsthand, where research is taking place at an institution. A good inspection experience will provide ample opportunity for researchers to describe what happens in their spaces and allow inspectors to ask questions of the research team. Inspections are also a chance for the librarian to meet individuals they may not traditionally be in contact with through outreach activities. This may include laboratory or facility managers, research technicians, and other research staff. These team members will have information needs related to training and keeping current on research practices. By participating in inspections, the librarian can make connections with these individuals, which may lead to later follow-up and assistance.

Training

Researchers are often unaware of what a good literature search looks like and are not familiar with the library resources available to them.[9] As a result, they may find that their protocol is not initially approved by the IACUC, and they are referred to the librarian for assistance.[10] In response, librarians can develop training for researchers on searching the literature for animal alternatives and meeting the requirements for a good search. In particular, new researchers or those submitting protocols for the first time should be offered consultations on searching for alternatives. Education about the search may also be incorporated into ongoing researcher training such as annual refreshers. Librarians may choose to develop resources such as guides and handouts that researchers can refer to as they prepare their protocols, with clear instructions on how to contact the librarian for assistance when needed. Training opportunities such as these provide the librarian with an opening to demonstrate to researchers their skills and the support they can offer, which may lead to further opportunities for collaboration.

As a member of the IACUC, the librarian can also provide education for other committee members on what a good alternatives search looks like and what should be included in an animal protocol. This can contribute to a more robust protocol review. The librarian can also contribute to keeping committee members current on new alternatives research and regulations by monitoring the literature for new publications. One such example is the literature search requirement itself, as there are recommendations to reduce administrative burden of researchers by eliminating the direct reporting of searches and instead ask researchers to provide a written description of their process to consider alternatives.[11] Relying on a librarian to keep the committee up to date on changes in the field such as this utilizes a librarian's skills to monitor the literature and remain current on trends, while helping the committee respond to regulatory developments in a timely manner.

Working with the Animal Care Team

The animal care team is an audience a librarian may not typically have contact with unless involved in IACUC meetings and other committee activities. Information needs of the animal care team include keeping current with research about animal health and welfare, access to resources for updating guidelines and standards of practice, and maintaining professional certifications. They are also often largely responsible for training researchers on proper animal handling and experimental techniques involving animals. Work with the IACUC provides an opportunity for the librarian to establish a partnership with animal care staff and explore ways to provide them with access to the research they need to remain current on best practices. The librarian may be asked to conduct searches for updating policies and practices, or for research on animals or techniques that may be new to the institution. The librarian should also consider resources that may be incorporated into training for researchers offered by the animal care staff, such as videos showing proper animal handling. A partnership with the animal care staff is highly beneficial as it helps maintain a high standard of care for the animals at an institution.

VALUE OF PARTNERING WITH THE IACUC

The role of librarians is evolving and, particularly in academic research settings, new opportunities are emerging that are driven by the need for more accountability by stakeholders and increased research funding.[12] Librarians are increasingly being called upon to demonstrate their value to their institution and partnering with an IACUC is a unique way to demonstrate the services of a librarian while directly contributing to research priorities. The benefits to the librarian are numerous and include improved working relationships with researchers and a better understanding of their work, increasing the librarian's visibility and opportunities for marketing their services, and enhancement of searching skills.[13] It also introduces the librarian to audiences that they may not typically interact with through normal channels of outreach and liaison work. Making these connections can result in further collaboration with research teams because the librarian is a known entity and has already shown how their work can fit into the research lifecycle.

Partnership with the librarian also provides important benefits to the IACUC itself. The literature search narrative is often the weakest part of animal protocols and librarian involvement strengthens the consideration of alternatives by researchers, increasing the number of animal protocols successfully approved by the committee.[14] In the greater scheme of animal welfare, it is important to remember that what can seem like numerous and complex administrative requirements are directly impacting the health and well-being of animals used in research. Review by a librarian to assure a proper search of the literature for alternatives can mean less pain and suffering for an animal used in an experiment, or even the difference between life and death.[15] A successful IACUC depends on these considerations to maintain an animal research program, which ultimately impacts research funding and success.

Working with an IACUC can be a time- and labor-intensive commitment for a librarian, but the benefits far outweigh the effort required. As a member of the committee, the librarian can provide unique knowledge and experience that contributes to improved animal protocols and consideration of alternatives, as well as better research practices and program guidelines. Through committee activities, the librarian will be exposed to new audiences and deepen relationships with the research community, while adding to their own skill set and demonstrating the value they add to the IACUC, the library, and the wider research community as a whole.

NOTES

1. Antony Brewerton, "Re-Skilling for Research: Investigating the Needs of Researchers and How Library Staff Can Best Support Them," *New Review of Academic Librarianship* 18, no. 1(2012): 96–110.

2. Anna Marie Johnson, "Connections, Conversations, and Visibility: How the Work of Academic Reference Liaison Librarians is Evolving," *Reference & User Services Quarterly* 58, no. 2(2018): 91–102.

3. Kara M. Whatley, "New Roles of Liaison Librarians: A Liaison's Persepective," *Research Library Issues* 265(2009): 29–32.

4. Susan C. Steelman and Sheila L. Thomas, "Academic Health Sciences Librarians' Contributions to Institutional Animal Care and Use Committees," *Journal of the Medical Library Association* 102, no. 3(2014): 215–19.

5. Animal Welfare Information Center (AWIC), "Meeting the Information Requirements of the Animal Welfare Act: A Workshop" (Beltsville, MD: United States Department of Agriculture, 2011).

6. Marina Chilov and Konstantina Matsoukas, "Library Support for an Academic Medical Center's Institutional Animal Care and Use Committee (IACUC)" (poster presented at the Medical Library Association Annual Conference, San Antonio, Texas, 2005).

7. Mary Auckland, *Re-Skilling for Research,* Research Libraries UK, January 2012.

8. Swapna Mohan, Robert W. Barbee, and Susan B. Silk, "Contribution of the Breadth and Depth of IACUC Membership to Experimental Design as a Factor in Research Reproducibility," *Journal of the American Association for Laboratory Animal Science* 57, no. 2(2018): 104–9.

9. John E. Osinski, "The Animal Welfare Act and Why It Matters to Librarians," *Reference Services Review* 39, no. 2(2011): 318–34.

10. Chilov and Matsoukas, "Library Support."

11. National Institutes of Health (NIH), Animal and Plant Health Inspection Service (APHIS), the United States Department of Agriculture (USDA), and the Food and Drug Administration (FDA) [the 21st Century Cures Act § 2034(d) Working Group], *Reducing Administrative Burden for Researchers: Animal Care and Use in Research*, published August 2019, https://olaw.nih.gov/sites/default/files/21CCA_final_report.pdf.

12. Johnson, "Connections, Conversations, and Visibility."

13. Steelman and Thomas, "Academic Health Sciences Librarians' Contributions."

14. Virgina A. Lingle, "The Health Sciences Library and the IACUC in Animal Research: Collaboration for More Effective Use of Electronic Resources," *Journal of Electronic Resources in Medical Libraries* 5, no. 3(2008): 243–59.

15. Osinski, "The Animal Welfare Act."

BIBLIOGRAPHY

Animal Welfare Information Center (AWIC). "Meeting the Information Requirements of the Animal Welfare Act: A Workshop." Beltsville, MD: United States Department of Agriculture, 2011.

Auckland, Mary. *Re-Skilling for Research*. Research Libraries UK, January 2012.

Brewerton, Antony. "Re-Skilling for Research: Investigating the Needs of Researchers and How Library Staff Can Best Support Them." *New Review of Academic Librarianship* 18 no. 1(2012): 96–110. doi: 10.1080/13614533.2012.665718.

Chilov, Marina, and Konstantina Matsoukas. "Library Support for an Academic Medical Center's Institutional Animal Care and Use Committee (IACUC)." Poster presented at the Medical Library Association Annual Conference, San Antonio, Texas, 2005.

Johnson, Anna Marie. "Connections, Conversations, and Visibility: How the Work of Academic Reference Liaison Librarians is Evolving." *Reference & User Services Quarterly* 58 no. 2(2018): 91–102.

Lingle, Virginia A. "The Health Sciences Library and the IACUC in Animal Research: Collaboration for More Effective Use of Electronic Resources." *Journal of Electronic Resources in Medical Libraries* 5, no. 3(2008): 243–59. doi: 10.1080/15424060802222414.

Mohan, Swapna, Robert W. Barbee, and Susan B. Silk. "Contribution of the Breadth and Depth of IACUC Membership to Experimental Design as a Factor in Research Reproducibility." *Journal of the American Association for Laboratory Animal Science* 57, no. 2(2018): 104–9.

National Institutes of Health (NIH), Animal and Plant Health Inspection Service (APHIS), the United States Department of Agriculture (USDA), and the Food and Drug Administration (FDA) [the 21st Century Cures Act § 2034(d) Working Group]. *Reducing Administrative Burden for Researchers: Animal Care and Use in*

Research. Published August 2019. https://olaw.nih.gov/sites/default/files/21CCA
_final_report.pdf.

Osinski, John E. "The Animal Welfare Act and Why It Matters to Librarians." *Reference Services Review* 39, no. 2(2011): 318–34. doi: 10.1108/00907321111135510.

Steelman, Susan C., and Sheila. L. Thomas. "Academic Health Sciences Librarians' Contributions to Institutional Animal Care and Use Committees." *Journals of the Medical Library Association* 102, no. 3(2014): 215–19. doi: 10.3163/1536-5050.102.3.014.

Whatley, Kara M. "New Roles of Liaison Librarians: A Liaison's Persepctive." *Research Library Issues* 269(2009): 29–32.

Chapter Eight

The Novice Investigator

Learning Responsible Research through Librarian-led Educational Efforts

Robin E. Champieux, Marijane K. White,
and Kristine M. Alpi

We gratefully acknowledge the contributions of Dr. Jean-Phillip Gourdine
to the development of the Rigor and Reproducibility nanocourse discussed
in this chapter.

This chapter describes librarian-led contributions to the research education
of graduate students and other investigators at two diverse institutions. We
situate our educational efforts within the larger context of training on all
aspects of responsible conduct of research (RCR) through a brief summary
of the published literature on RCR training, as well as a report of IACUC-
and IRB-related learning resources provided by libraries. We share specific
examples of our collaborations with colleagues to deliver RCR training,
including efforts to support research trainees in achieving research reproducibility, recognizing its essentialness to the advancement of knowledge and
responsible research.

LITERATURE REVIEW

In the United States, research trainees supported by the National Institutes of
Health (NIH) and the National Science Foundation (NSF) must receive responsible conduct of research (RCR) training. In the case of the NIH, specific
training requirements address subject matter, instructional format, duration,
frequency, and faculty involvement.[1] While not specified in its policy, in
2008, the NSF sponsored a workshop to elucidate and build consensus around
best practices for RCR education. The recommendations are similar to the
NIH requirements and have been promoted to educational institutions in order to guide the design of their training programs.[2] Both agencies underscore

the importance of interactive, mixed-format instruction involving faculty that takes place across the duration of the training experience.

When we searched PubMed in July 2020, the literature contained several articles describing educational approaches to RCR, including processes and requirements specific to institutional review boards (IRBs) and Institutional Animal Care and Use Committees (IACUCs). The University of Pittsburgh developed and deployed online modules addressing research integrity, human subjects research, laboratory animal research, and conflict of interest.[3] Smaller-scale education efforts included mock IRB,[4] role-play,[5] and an annual group training for doctor of nursing practice (DNP) students at the University of Iowa College of Nursing using a decision analysis tool to address determination of whether or not a project constituted research.[6] One of the few studies following up with participants after IRB-related training came from the University of Ibadan, Nigeria, which reported that knowledge about ethics and IRB function was retained by 61 percent of attendees who completed the posttest at one month follow-up.[7] While IACUC educational studies are uncommon, Choe et al. describes training provided by agencies in Korea and suggests that curricula should vary based on the roles, backgrounds, and needs of the individual trainees, and that government authorities can encourage improvement of training curricula.[8]

Several studies examined the existence, design, and scope of RCR training across institutions through the lens of NIH and NSF training requirements or guidelines. For example, Phillips et al. analyzed the training plans of 103 institutions with an R1 Carnegie classification (meaning a "research university with very high research activity") and found that the majority provided instruction exclusively online and heavily relied on the Collaborative Institutional Training Initiative (CITI).[9] They concluded most publicly available plans did not meet the best practices articulated in the NSF workshop.[10] Similarly, Resnick and Dinse found that only 51 percent of the U.S. research institutions surveyed required RCR training for individuals not explicitly subject to the NIH's requirements. They encouraged institutions to expand their training programs to advance research integrity.[11] Finally, DuBois et al. found no consistent approach to RCR training among the institutions with a Clinical and Translational Science Award (CTSA), noting the lack of curriculum specific to translational research.[12] They highlighted the tension between their finding and expert opinions on best practices for RCR such as content and educational methods specific to discipline and career stage.[13]

Library and librarian contributions to formal RCR education programs are not well documented in the biomedical literature. Focusing the aforementioned search to include articles on research ethics education and libraries or librarians retrieved only one result, which was not specifically relevant.

However, a 2013 survey documented the role of thirty-six Association for Research Libraries (ARL) member libraries in providing RCR training.[14] The survey results provide a picture of the ways in which librarians are "promoting the responsible conduct of research through a variety of supplemental guides and training."[15] Most survey respondents (77 percent) offered some RCR-related training; the most prominent topics included citation practices and management, plagiarism, responsible authorship, and data management. Course-based instruction sessions (89 percent) and face-to-face workshops (83 percent) were the most common delivery formats, followed by online guides and tutorials. In contrast to many of the studies of RCR training described above, most library survey respondents described training approaches tailored to specific audiences. While not addressed in the 2013 survey, ARL libraries are increasingly promoting electronic laboratory notebooks (ELN) and providing training about the role of ELN in improving documentation (e.g., http://ebling.library.wisc.edu/2018/07/19/electronic-lab-notebooks-all -invited/).

A few studies detail library contributions to RCR through research data services. A 2013 survey of seventy-three ARL member libraries found that 74 percent offered some kind of data management services, defined as "providing information, consulting, training or active involvement in: data management planning, data management guidance during research (e.g., advice on data storage or file security), research documentation and metadata, research data sharing and curation (selection, preservation, archiving, citation) of completed projects and published data."[16] Yoon and Schultz examined the landscape of academic library data management services through a content analysis of 185 library websites.[17] They identified an array of services, including data deposition (60 percent), data management planning (41 percent), data consultations (38 percent), and data publishing and sharing (28 percent). Less than half of the library websites (35 percent) included descriptions of educational programs, but those that did offered classes and workshops on data management, sharing and reuse, and the use of specific methodologies or data types.

RCR includes competencies and practices aligned with core library services and librarian expertise, namely the organization, description, dissemination, and reuse of information. The importance of this knowledge and its ethical practice is clearly relevant to authorship, publication, and data management, which the NIH RCR training requirements include as subject requirements.[18] DuBois and Antes describe the five dimensions of research ethics, which are normative ethics, compliance, rigor and reproducibility, social value, and workplace relationships.[19] They regard appropriate dissemination as foundational to the social value of research and their description of rigor and reproducibility stresses properly documented and accessible data.[20] Similarly, the

National Academies of Sciences, Engineering, and Medicine's 2019 report, *Reproducibility and Replicability in Science*, emphasizes clear and complete documentation and data sharing, and contextualizes these and other practices as fundamental to modern, responsible research.[21] Since the literature demonstrates few cases of formal library or librarian involvement in RCR training, an understanding of library contributions to RCR education can proceed from the perspective of the information, services, and instructional efforts that speak to these components of responsible and ethical research.

TEACHING SEARCH DOCUMENTATION AS AN ELEMENT OF RESPONSIBLE CONDUCT OF RESEARCH

Librarians may not be traditionally viewed as having expertise in RCR.[22] When offering to contribute to a portfolio of RCR training at a large land-grant university library in the Southeastern United States for new investigators and graduate students, we chose to develop content related to how searching contributes to IRB and IACUC proposals. We felt this was an area where librarians would be recognized as experts and be able to continue to work with trainees after the workshop. The library-based Professional Development Seminar titled *How and Why to Document Searches in the Sciences: A Hands-On Workshop* described here was advertised both as part of the university's RCR program on a website geared toward letting graduate students and postdoctoral associates know of research training opportunities and on the library website as a general library workshop. The ninety-minute RCR offering in September 2010 involved nine graduate students from diverse departments. The second offering in November 2010 was the standard fifty-minute workshop in the library graduate student workshop series and was attended by five graduate students and two library staff. Both were held in the main library's computer lab.

The goals of the workshop were as follows:

1. improved clarity of methods sections in proposals and in manuscripts
2. enhanced communication with advisors and reviewers about one's exploration of relevant literature
3. reproducibility by others where the search or analysis strategy is part or all of the experiment

The workshop materials covered several complementary documentation strategies. First was recording search elements in a log or laboratory notebook, like any other aspect of research effort. Using personal accounts to save searches in databases with system storage options was discussed along with the potential to overwrite dates searched or last updated, and dealing with default parameters that change over time. This led to sharing options for capturing searches and results at the time searched, and then how to organize and store these searches and their outputs (e.g. results, chosen citations, full text of articles). The last page of the handout was a sample search log form (see figure 8.1) with the elements and strategies for file management, which was also available as a Word document on the library website.

In the training, we delved into three key elements of searching and why they are crucial to documentation: (1) date the search was performed, (2) databases searched, and (3) time periods covered or date limits applied in the search. We discussed the date of search as a legal issue for intellectual

Project: _____

Date Searched: _____   Searcher: _____

System: _____

Database(s): _____

(Include name, year segments if applicable)

    Search history attached or saved (location/filename: _____)

      Update search saved or ◻ alert enabled online [in case you decide to edit]

o Username and/or password: _____

    Relevant results saved in bibliographic management software

o Username and/or password: _____

o Folder: _____

    Related articles searched; specify papers with PMID or Accession #:

_____

_____

Search Terms Used: (include truncation, Boolean operators, nesting)

Search Limits Applied: (Language, Publication Type, Date)

Figure 8.1. Sample Search Log.
Robin E. Champieux, Marijane K. White, and Kristine M. Alpi

property or capturing known risks at the time of the study based on whether knowledge was "published" and "available" at the documented time of search. The PubMed/MEDLINE database exemplified the need to document the time period since it initially covered 1966 onward but later added earlier year citations from the OLDMEDLINE database.[23] The workshop benefited from having students from multiple disciplines discuss the value of following the literature during a project beyond the extent of responsibility investigators have to monitor for updates. Specific examples include becoming aware of new information on risks or benefits that may lead you to want to adjust your protocol with the IRB or IACUC or, if the data has already been collected, addressing new information in your discussion as you write up your publication. Techniques such as manually recreating searches or setting up search alerts with reasonable frequencies were discussed and again varied by discipline. We revisited this discussion within the IRB case example specific to addressing the IRB annual review for multi-year projects, as the form at our institution asked about new relevant literature or updates.

Participants reflected on the databases they typically search and were asked to use the product information or the library's database description to identify what years were covered. We discussed that for products like Web of Science/Web of Knowledge, the time frames and constituent databases for an All Databases search vary by the institutional subscription, so what they use as a student may differ from what they have at their postdoctoral or faculty position. Further, content in databases like Google Scholar are constantly changing, so best practice is to capture the pages of results actually perused as PDFs or screen captures. Students and instructors shared personal practices of searching and ideas for enhancing their approaches to facilitate future documentation. Most had no personal documentation practices. They indicated they would only have documented it if required for an assigned task and would have followed the instructions provided.

The only required search documentation we shared was the required component of the IACUC Alternatives search. None of the first workshop students were planning to work with animals and, therefore, were not familiar with the IACUC form. However, they agreed that as graduate students who are talking with their mentors about what they did or did not find in the literature, it could be helpful to document the keywords or subject headings used and how they were combined. We demonstrated documenting nesting, truncation, Boolean operator use, and using generic terms and trade names with this example: (game* or gaming or Halo or Wii). We also talked about documenting article citations or unique IDs followed for "pearl-growing," "snowball searching," "cited by," and "citing references" strategies. In addition to documenting the search in order to reproduce what it retrieved, we delved into the larger issue of what it means to have retrieved and reviewed the literature. An investigator is responsible for being familiar with the information in all the retrieved search results even if they chose not to read most of the abstracts. For search results presented in order of relevance, no one was in the habit of documenting where they stopped reviewing the results even though this is an issue not just for database results, but also Google Scholar/ web results where the position of results changes over time as new content is added or algorithms change.

Participant thoughts on ideal storage for their searches varied most by personal approach to organization and file management. The options depended on how researchers kept their lab notebooks and whether they thought they would need access to the materials as they transitioned institutions from graduate school to a post-doc to a professional position. The first was to save search histories and create alerts in the databases in which they were created, including ensuring that the original date searched and subsequent search dates and parameters were recorded. An option for web-based searches (e.g., Google Scholar) was to print web search history and results pages to PDF (which includes the system date) and store the PDFs as documents in a citation management tool or other storage programs. For those that used an online lab notebook or maintained files in a shared system like Google Docs, Box, or Dropbox, one easy option was copying and pasting search history into a search log/file maintained digitally for that project. Another was to provide file information in their lab notebook or project tracking system. For print-reliant researchers, the option of printing the search history and including it on dated and numbered pages in the notebook was a possibility. We also discussed combination approaches.

We then went over four case examples to demonstrate why documenting searches is an important part of one's research methods. The first two case examples were focused on critically reviewing methods from published papers and the last two focused on the IACUC and the IRB forms. For this chapter, details are only provided for the IACUC and IRB examples.

Case Examples

1. *in silico* bioinformatics: showing how BLAST search described in *Cell* (doi: 10.1016/j.cell.2010.01.032) and *Science* (doi:10.1126/science.1160619) articles compared with instructions for authors requirements
2. systematic review/meta-analysis: documenting search is part of method and some fields like medicine have specific guidelines like PRISMA; compare software engineering systematic review in *Information and*

Software Technology (doi:10.1016/j.infsof.2010.05.004) with *JAMA* example using PRISMA (doi:10.1001/jama.299.16.1937)
3. animal use studies—show required search elements of IACUC application form
4. human subjects research—show IRB continuing review form and discuss investigator documentation of search in an ethics case from Johns Hopkins University

ANIMAL USE STUDIES

This section focused on what the institutional IACUC submission form required as search documentation (shown below). Participants compared these requirements to their search logs to ensure that they would be capturing all of the required elements.

There must be a written narrative description of the methods and sources that were consulted to determine the availability of alternatives (reduction, refinements, replacement).

1. Please respond to items a–d regarding your literature search for alternatives:
   a. Which databases were searched?
   b. Indicate the range of dates searched within the database(s) (i.e., 1900–2009, or 1987–present):
   c. Keywords should include those likely to yield information on alternatives to the potentially painful or distressful procedures or conditions that are part of this protocol.
   d. List keywords used:
   e. Provide the most recent date that a full search was performed:

We discussed that the use of "keywords" on the form meant not only keywords, but also subject headings/controlled vocabulary terms with their appropriate syntax. We also discussed why showing the combination of search terms used, not just a list of them, was important for being able to reproduce the search.

HUMAN SUBJECTS RESEARCH

This section focused on the search elements of a widely publicized IRB investigation at Johns Hopkins of the case of Ellen Roche, who died in June 2001 after an asthma study. The search to ascertain the drug's adverse effects was documented as part of the investigation report, allowing students new to

IRB applications and human participant research a chance to see how their search documentation may be scrutinized. The original study investigator's search focused on a limited number of resources, including PubMed, which at that time was searchable only back to 1966, and the relevant articles warning of risk of lung damage had been published in the 1950s with citations in subsequent publications. However, the PubMed original search included a 1970 review article that listed six references from the 1950s if the investigator had read it. This allowed the workshop participants to revisit the question of not just retrieval but how deeply to have read or reviewed the retrieved literature.

At the time of the workshop, the Johns Hopkins Internal Investigation Report was online, and we reviewed that original report. The investigator said that he had performed a standard PubMed search for potential hexamethonium toxicity and consulted standard, current edition textbooks of pharmacology and pulmonary medicine before submitting the application to the IRB. The details of the search documentation for the PubMed component state,

> During PubMed searches, "hexamethonium inhalation lung injury" gave 0 hits, "hexamethonium inhalation" gave 42 hits (but none referring to pulmonary toxicity), "hexamethonium lung" yielded 3 useful articles, "hexamethonium lung toxicity" gave 4 hits, but 0 useful articles, "hexamethonium lung hypersensitivity" gave 16 hits with 3 useful articles, and "hexamethonium lung fibrosis" gave 3 hits and 2 useful articles.

We discussed how the search strategies and documentation here compared with our earlier discussions about required elements and the challenges of looking back in time at changes in resources. We noted that these were not part of the IRB-required paperwork; they were part of the investigator's research documentation. This led to our examination of what was required by the IRB at our institution at the time. We first looked at the submission for new studies, which requires a statement of potential risks and potential benefits and found it to be silent on how investigators should document their search for risks. The form for continuing review of existing studies—typically done annually for multi-year studies—was more explicit. It asks the following:

> Is there any new information since the last IRB review that might impact the risks vs. benefits of the research? If so, please submit a summary of any recent literature, findings, or other relevant information, especially information about risks associated with the research.

At the end, we discussed the limited role of the IRB in considering literature searches based on the 2007 guidance from the Office for Human

Research Protections, Department of Health and Human Services,[24] which was in effect at the time of the course and stated the following:

> *What is the role of the IRB in literature searches at continuing review?* The regulations do not state or even suggest that the IRB is required to perform or validate a review of the literature. Reviewing the literature is a scientific activity and as such is the responsibility of the investigator—the IRB receives the results of the review.

This guidance was updated in November 2010 after the second workshop was taught, and this reinforces that librarians teaching in RCR must make sure that they are updating training or materials to reflect the most current guidance, or clearly post on training materials the date of guidance on which the materials were developed so that trainees do not inadvertently rely on outdated content. We chose not to offer this course as a workshop after the second offering because the audience's informal feedback was that it would be better to learn about this individually or with a group of peers doing the same type of research rather than learning about this in the context of types of research that they would never do; for example, those who do computational or bench research were not interested in the animal or human participants documentation needs, and those doing animal research were not interested in human and vice versa. The knowledge from developing this course was therefore further shared as librarians consulted individually or in small groups with graduate students, post-doctoral fellows, and new investigators rather than coordinating discipline-specific workshops on this topic.

LIBRARIANS TEACHING RIGOR, REPRODUCIBILITY, AND RESEARCH DATA MANAGEMENT

As an academic health sciences center library in the Pacific Northwest, our library faculty's instructional involvement in the education of basic science graduate students includes guest lectures and library-led workshops. This portfolio of educational contributions leverages partnerships with subject experts and has evolved to address learning goals and needs not necessarily met by the formal curriculum. The topics addressed include literature searching, reference management, publishing and scientific communication, open science, and various data-focused topics, including data documentation, computer programming, and data visualization.

In 2018, we approached the Program in Molecular and Cellular Biosciences with a proposal to reconceive the instruction of its Rigor and Reproducibility nanocourse, which was a 0.5 credit course designed, in part, to meet the NIH's RCR training requirements for learners supported by the Program in Enhanced Research Training (PERT). PERT is an institutional research training grant from the NIH for second-year graduate students focused on developing competencies needed for successful careers in biomedical research. The course, which was graded pass/no pass and not included in GPA calculations, was required for PERT students and was also available as an elective to all graduate students in the School of Medicine. This was an exciting opportunity to position the library as a partner in promoting research reproducibility and to showcase our related services and expertise, most specifically research data services. While we had contributed to and had been responsible for portions of for-credit courses in the past, this was the first time the library would be the primary party responsible for designing and teaching its own implementation of a for-credit course at our institution.

The course met in the spring of 2019 for a total of six instructional hours over five weeks in a conference room on campus and enrolled five PhD students from the PERT program. Course materials were based on the NIH's Rigor and Reproducibility Training Modules, a four-part course oriented around a set of short videos and discussion questions.[25] The four NIH modules are (1) Lack of Transparency, (2) Masking and Randomization, (3) Biological and Technical Replicates, and (4) Sample Size, Outliers, and Exclusion Criteria. We added a fifth module on RDM, based on NYU's Research Data Management Teaching Toolkit.[26] We also augmented the course materials with additional resources, including optional readings, tools for reproducibility, a checklist of best practices for preclinical research based on Steward and Balice-Gordon's *Rigor or Mortis: Best Practices for Preclinical Research in Neuroscience,*[27] and case studies, which we used for in-class exercises. We documented all of the articles referenced in the source materials and our own curated resources in an extensive course bibliography to which we hoped our students would refer as they pursued rigor and reproducibility in their research careers.

The learning objectives for each module of the nanocourse were as follows:

Week 1: Introduction and Lack of Transparency
- Understand factors that contribute to reproducibility issues
- Understand the elements of rigorous study design
- Prepare students to enhance experiments in terms of design, execution, analysis, interpretation, and reporting

Week 2: Research Data Management
- Recognize current and upcoming requirements mandating the management and sharing of research data
- Understand best practices for creating and documenting file names, variables, and workflows

- Identify appropriate methods for storing and preserving research data
- Locate appropriate standards, if any, and recognize their value for research
- Evaluate repositories and determine best data sharing options

Week 3: Masking and Randomization
- Understand what masking and randomization are
- Describe why masking and randomization are important techniques to use in research
- Understand what activities can be masked and different types of masking
- Describe and apply different randomization techniques

Week 4: Biological and Technical Replicates
- Understand the process of replication
- Describe the difference between biological replicates and technical replicates
- Explain practical examples of replicates

Week 5: Sample Size, Outliers, and Exclusion Criteria
- Identify how to approach sample size and power calculations
- Describe essential activities for characterizing data
- Illustrate techniques for identifying and investigating outliers in data

We originally intended to teach the research data management module during the last week of the course, but moved it to the second week due to a co-facilitator's scheduling conflict. The standalone nature of the RDM module meant it could be taught at any point in the nanocourse, but it seemed to fit well after the introductory Lack of Transparency module, which provides an overview of some of the issues RDM aims to address.

We knew in advance of proposing to teach the course that some of the material in the NIH training modules were beyond our expertise, as we lacked our own scientific training and could not speak with authority on topics such as biological and technical replicates. Therefore, as with previous workshops, we used our knowledge of and relationships with subject matter experts to address these topics and associated learning outcomes. For example, we asked Jean-Phillip Gourdine, PhD, a glycobiologist working for the library on a campus-wide data needs assessment, to co-facilitate the nanocourse. We knew that Dr. Gourdine's scientific knowledge would not only fill in our instructional gaps but that he could offer framing and experience throughout the course relevant to our students' experiences and goals. We also invited an assistant professor of biostatistics in the School of Public Health to reprise a presentation she had recently given on approaching sample size and power calculations for the Sample Size, Outliers, and Exclusion Criteria module.

A typical class meeting began with viewing a NIH module video followed by a group discussion. Rather than instructor-led discussion, we used the Cephalonian Method, an active learning technique where questions are printed on cards and passed out to students before the exchange begins.[28] Students read questions to the group when the number on the back of the card is called by an instructor. We used the discussion questions from the NIH materials as a starting point for our question cards, editing them to be more appropriate to our audience and creating our own set of questions for the RDM module. This was an easy and effective way to increase student engagement. The discussion was most often followed by a lecture from a course facilitator or guest speaker, and we wrapped up with a case study exercise. In the first meeting of the course, Dr. Gourdine asked our students about their background and research focus, and we used this information to select case study examples that would reflect their research interests. The RDM module had a different flow, as it was more lecture-focused and did not involve any case study exercises, but it also included a cartoon video created by NYU's Health Sciences Library describing a worst-case data management scenario, which our students appeared to enjoy. Our complete course materials—including our syllabus, lecture slides, course bibliography, and Cephalonian Method question cards—have been published openly online.[29]

To assess our learners' engagement and refine subsequent lectures, we asked the students to anonymously answer an optional Qualtrics survey, containing two open-ended questions, at the end of each class meeting, except for the last meeting, where we instead asked students to evaluate the course as a whole using a Qualtrics survey based on the six-question evaluation form included with the NYU RDM Teaching Toolkit.[30]

The two-question survey asked the following:

1. What do you want to take away from today's class?
2. What topics discussed in today's class would you like to learn more about?

The first question encouraged reflection on the student's part and the second was designed to help us improve the course content. Each week, we received at least two responses to this survey resulting in eleven total responses during the course. Answers to the first question were primarily about the reproducibility practices, tools, and techniques students felt they could implement in their work, but one notable exception was a student whose take away from the first week was that "most science is no longer reproducible, and that needs to change." All of the responses to the second question in the first week expressed interest in learning more about topics that would be covered in later course modules. In subsequent weeks, the students expressed a desire to go into more detail, to see more real-world examples, and to learn how to use the tools and techniques addressed in the lectures. This revealed potential topics for future library workshops, such as focused sessions on some of the tools mentioned in the RDM module. While only one of the five students completed the final course evaluation, the responses to all the surveys were overwhelmingly positive.

Due to both student and graduate program faculty reception to the course, the library has been invited to teach an expanded version beginning Fall of 2021, as part of the university's new Graduate Program in Biomedical Sciences (PBMS). PBMS combines several existing PhD programs into one program and is the result of several years of assessment and planning. The new library-led Rigor and Reproducibility course will be 1-credit hour (twelve instructional hours) and required for all students, regardless of their NIH-support status. The library has also been asked to design and teach several additional for-credit courses in the new program. While not specifically intended to address RCR training requirements, the classes are focused on knowledge and practices essential to responsible research, such as searching, reference management, and scientific communication, and we involve more library faculty and their expertise in the teaching process.

Our experience with the Rigor and Reproducibility course increased our instructional capacity and approach in surprising ways. The RDM module we built and added to the original course can be taught as a standalone session. We have presented it to students in our MD/PhD program and components of it have been incorporated into various professional development workshops. Because the expanded course will include twelve instructional hours, we plan to develop an additional module on our institution's resources for rigor and reproducibility, which we expect to be able to use similarly. Moreover, our successful use of the Cephalonian Method has inspired us to more consistently incorporate this and other active learning techniques into our instruction.

Finally, our involvement with this course and other workshops that address practices foundational to responsible and data-intensive research motivates us to describe these skills in ways that facilitate learner assessment and the discovery of educational opportunities. We successfully pursued a Research Data Engagement Award in 2019 from the Network of the National Library of Medicine, Pacific Northwest Region, and are developing a framework for describing these skills, including competencies related to ethics, communicating results, and rigor and reproducibility. We believe the framework will help learners assess their current knowledge and navigate learning opportunities the library and others offer. The framework will be published under a Creative Commons license, so others can adapt and apply it.

CONCLUSION

Institutions should deliver long-term, deliberate, and customized RCR training in order to support an ethical culture of scientific investigation.[31] As exemplified by the cases and literature we described, libraries can (and do!) make important contributions to building novice researchers' understanding of practices foundational to RCR. There is also evidence that libraries are experiencing and applying best practices for delivering RCR training. For example, the students in our literature search documentation workshop reported that a discipline-specific approach would have been more valuable. Likewise, the learners in our Rigor and Reproducibility course likely benefited from being at the same career stage and in the same PhD program. By supplementing the NIH's training materials with readings and case studies that spoke to learners' research experiences and interests, we customized the course to their disciplinary perspective in a way that could be replicated for other audiences. While the new PhD program in which we will teach this course come fall 2021 is multi-disciplinary, we will have the opportunity to engage with students at the same career stage at more points along their discipline-specific educational experience.

While library and librarian contributions to RCR training are not well documented in the literature, our current and potential involvement becomes clearer when examined through an understanding of what responsible research encompasses. This can be encouraging from a capacity perspective; for many libraries, it could be relatively easy to reframe existing services and expertise to make their relationship to RCR competencies more explicit. Yoon and Shultz contend that academic libraries are uniquely positioned to respond to their community's research data management and stewardship needs because of their expertise and deep relationships with campus-wide stakeholders.[32] The same could be said of our positioning to contribute to RCR training and research integrity more broadly. Libraries wishing to grow and receive credit for their contributions to RCR education may find it helpful to specifically promote their expertise in ways that speak to the requirements and best practices institutions are striving to meet. In this way, we will continue to contribute to the overall improvement of RCR education.

NOTES

1. National Institutes of Health, "NOT-OD-10-019: Update on the Requirement for Instruction in the Responsible Conduct of Research," November 24, 2009, https://grants.nih.gov/grants/guide/notice-files/NOT-OD-10-019.html.

2. Carol R. Arenberg and Rachel Hollander, *Ethics Education and Scientific and Engineering Research What's Been Learned? What Should Be Done?: Summary of a Workshop* (Washington, DC: National Academies Press, 2009), https://www.nap.edu/catalog/12695/ethics-education-and-scientific-and-engineering-research-whats-been-learned. Trisha Phillips, Franchesca Nestor, Gillian Beach, and Elizabeth Heitman, "America COMPETES at 5 Years: An Analysis of Research-Intensive Universities' RCR Training Plans," *Science and Engineering Ethics* 24, no. 1 (2018): 227–49, https://doi.org/10.1007/s11948-017-9883-5.

3. Barbara E. Barnes, Charles P. Friedman, Jerome L. Rosenberg, Joanne Russell, Ari Beedle, and Arthur S. Levine, "Creating an Infrastructure for Training in the Responsible Conduct of Research: The University of Pittsburgh's Experience," *Academic Medicine* 81, no. 2 (February 2006): 119–27.

4. Jean Dowling Dols, Mary M. Hoke, and Maureen L. Rauschhuber, "Mock Institutional Review Board: Promoting Analytical and Reasoning Skills in Research Ethics," *Nurse Educator* 42, no. 6 (December 2017): E4, https://doi.org/10.1097/NNE.0000000000000377.

5. Ralph L. Rosnow, "Teaching Research Ethics through Role-Play and Discussion," *Teaching of Psychology* 17, no. 3 (1990): 179–81, https://doi.org/10.1207/s15328023top1703_10.

6. Jan M. Foote, Virginia Conley, Janet K. Williams, Ann Marie McCarthy, and Michele Countryman, "Academic and Institutional Review Board Collaboration to Ensure Ethical Conduct of Doctor of Nursing Practice Projects," *Journal of Nursing Education* 54, no. 7 (June 25, 2015): 372–77, https://doi.org/10.3928/01484834-20150617-03.

7. Ademola J. Ajuwon and Nancy Kass, "Outcome of a Research Ethics Training Workshop among Clinicians and Scientists in a Nigerian University," *BMC Medical Ethics* 9, no. 1 (January 24, 2008): 1, https://doi.org/10.1186/1472-6939-9-1.

8. Byung In Choe and Gwi Hyang Lee, "Individual and Collective Responsibility to Enhance Regulatory Compliance of the Three Rs," *BMB Reports* 47, no. 4 (2014): 179–83, https://doi.org/10.5483/BMBRep.2014.47.4.049.

9. Phillips et al., "America COMPETES at 5 Years."

10. Ibid.

11. David B. Resnik and Gregg E. Dinse, "Do U.S. Research Institutions Meet or Exceed Federal Mandates for Instruction in Responsible Conduct of Research? A National Survey," *Academic Medicine: Journal of the Association of American Medical Colleges* 87, no. 9 (September 2012): 1237–42, https://doi.org/10.1097/ACM.0b013e318260fe5c.

12. James M. DuBois, Debie A. Schilling, Elizabeth Heitman, Nicholas H. Steneck, and Alexander A. Kon, "Instruction in the Responsible Conduct of Research: An Inventory of Programs and Materials within CTSAs," *Clinical and Translational Science* 3, no. 3 (June 2010): 109–11, https://doi.org/10.1111/j.1752-8062.2010.00193.x.

13. Ibid.

14. Michelle Leonard and Denise Beaubien Bennett, "Responsible Conduct of Research Training, SPEC Kit 336 (September 2013)," September 16, 2013, https://publications.arl.org/Responsible-Conduct-Research-Training-SPEC-Kit-336.

15. Ibid.

16. David Fearon Jr., Betsy Gunia, Sherry Lake, Barbara E. Pralle, and Andrew L. Sallans, "Research Data Management Services, SPEC Kit 334 (July 2013)," August 1, 2013, https://publications.arl.org/Research-Data-Management-Services-SPEC-Kit-334.

17. Ayoung Yoon and Teresa Schultz, "Research Data Management Services in Academic Libraries in the US: A Content Analysis of Libraries' Websites," *College & Research Libraries* 78, no. 7(2017): 920, https://doi.org/10.5860/crl.78.7.920.

18. National Institutes of Health, "NOT-OD-10-019: Update on the Requirement for Instruction in the Responsible Conduct of Research."

19. James M. DuBois and Alison L. Antes, "Five Dimensions of Research Ethics: A Stakeholder Framework for Creating a Climate of Research Integrity," *Academic Medicine: Journal of the Association of American Medical Colleges* 93, no. 4 (2018): 550–55, https://doi.org/10.1097/ACM.0000000000001966.

20. Ibid.

21. Engineering National Academies of Sciences, *Reproducibility and Replicability in Science* (Washington, DC: National Academies Press, 2019), https://www.nap.edu/catalog/25303/reproducibility-and-replicability-in-science.

22. Jessica Adamick, "Ethics Day: Engaging Librarians in the Responsible Conduct of Research," Ethics in Science and Engineering National Clearinghouse Beta Project (University of Massachusetts, Amherst, October 8, 2010), https://www.umass.edu/sts/digitallibrary/pubspres/adamick_ethicsday.pdf.

23. National Library of Medicine, "OLDMEDLINE Data" (U.S. National Library of Medicine, December 29, 2017), https://www.nlm.nih.gov/databases/databases_oldmedline.html.

24. Office for Human Research Protections, Department of Health and Human Services, "Guidance on Continuing Review–January 2007," Text, HHS.gov, April 7, 2016, https://www.hhs.gov/ohrp/regulations-and-policy/guidance/continuing-review-january-2007/index.html.

25. NIH Office of the Director, "NIGMS Clearinghouse for Training Modules to Enhance Data Reproducibility: NIH Rigor and Reproducibility Training Modules," 2015, https://www.nigms.nih.gov/training/pages/clearinghouse-for-training-modules-to-enhance-data-reproducibility.aspx#nih-rigor-and-reproducibility-training-modules.

26. Kevin Read and Alisa Surkis, "Research Data Management Teaching Toolkit," February 1, 2018, https://doi.org/10.6084/m9.figshare.5042998.v6.

27. Oswald Steward and Rita Balice-Gordon, "Rigor or Mortis: Best Practices for Preclinical Research in Neuroscience," *Neuron* 84, no. 3 (November 5, 2014): 572–81, https://doi.org/10.1016/j.neuron.2014.10.042.

28. Nigel Morgan and Linda Davies, "How Cephalonia Can Conquer the World (Or at the Very Least, Your Students!): A Library Orientation Case Study from Cardiff University," in *Practical Pedagogy for Library Instructors: 17 Innovative Strate-*

gies to Improve Student Learning (Chicago, IL: Association of College and Research Libraries, 2008), 20.

29. Marijane White and Jean-Philippe Gourdine, "Data Rigor and Reproducibility," July 29, 2020, https://doi.org/10.17605/OSF.IO/TEGY9.

30. Read and Surkis, "Research Data Management Teaching Toolkit."

31. Resnik and Dinse, "Do U.S. Research Institutions Meet or Exceed Federal Mandates for Instruction in Responsible Conduct of Research?"

32. Yoon and Schultz, "Research Data Management Services in Academic Libraries in the US."

BIBLIOGRAPHY

Adamick, Jessica. "Ethics Day: Engaging Librarians in the Responsible Conduct of Research," Ethics in Science and Engineering National Clearinghouse Beta Project (University of Massachusetts, Amherst, October 8, 2010). https://www.umass.edu/sts/digitallibrary/pubspres/adamick_ethicsday.pdf.

Ajuwon, Ademola J., and Nancy Kass. "Outcome of a Research Ethics Training Workshop among Clinicians and Scientists in a Nigerian University." *BMC Medical Ethics* 9, no. 1 (January 24, 2008): 1–9. https://doi.org/10.1186/1472-6939-9-1.

Arenberg, Carol R., and Rachelle Hollander. *Ethics Education and Scientific and Engineering Research What's Been Learned? What Should Be Done? Summary of a Workshop.* Washington, DC: National Academies Press, 2009. https://www.nap.edu/catalog/12695/ethics-education-and-scientific-and-engineering-research-whats-been-learned.

Barnes, Barbara E., Charles P. Friedman, Jerome L. Rosenberg, Joanne Russell, Ari Beedle, and Arthur S. Levine. "Creating an Infrastructure for Training in the Responsible Conduct of Research: The University of Pittsburgh's Experience." *Academic Medicine* 81, no. 2 (February 2006): 119–27.

Choe, Byung In, and Gwi Hyang Lee. "Individual and Collective Responsibility to Enhance Regulatory Compliance of the Three Rs." *BMB Reports* 47, no. 4 (2014): 179–83, https://doi.org/10.5483/BMBRep.2014.47.4.049.

Dols, Jean Dowling, Mary M. Hoke, and Maureen L. Rauschhuber. "Mock Institutional Review Board: Promoting Analytical and Reasoning Skills in Research Ethics." *Nurse Educator* 42, no. 6 (December 2017): E4–E8, https://doi.org/10.1097/NNE.0000000000000377.

DuBois, James M., and Alison L. Antes. "Five Dimensions of Research Ethics: A Stakeholder Framework for Creating a Climate of Research Integrity." *Academic Medicine: Journal of the Association of American Medical Colleges* 93, no. 4 (2018): 550–55, https://doi.org/10.1097/ACM.0000000000001966.

DuBois, James M., Debie A. Schilling, Elizabeth Heitman, Nicholas H. Steneck, and Alexander A. Kon. "Instruction in the Responsible Conduct of Research: An Inventory of Programs and Materials within CTSAs." *Clinical and Translational Science* 3, no. 3 (June 2010): 109–11, https://doi.org/10.1111/j.1752-8062.2010.00193.x.

Engineering National Academies of Sciences. *Reproducibility and Replicability in Science*. Washington, DC: National Academies Press, 2019. https://www.nap.edu/catalog/25303/reproducibility-and-replicability-in-science.

Fearon Jr., David, Betsy Gunia, Sherry Lake, Barbara E. Pralle, and Andrew L. Sallans. "Research Data Management Services, SPEC Kit 334 (July 2013)." Published August 1, 2013. https://publications.arl.org/Research-Data-Management-Services-SPEC-Kit-334.

Foote, Jan M., Virginia Conley, Janet K. Williams, Ann Marie McCarthy, and Michele Countryman. "Academic and Institutional Review Board Collaboration to Ensure Ethical Conduct of Doctor of Nursing Practice Projects." *Journal of Nursing Education* 54, no. 7 (June 25, 2015): 372–77. https://doi.org/10.3928/01484834-20150617-03.

Leonard, Michelle, and Denise Beaubien Bennett. "Responsible Conduct of Research Training, SPEC Kit 336 (September 2013)." Published September 16, 2013. https://publications.arl.org/Responsible-Conduct-Research-Training-SPEC-Kit-336.

Morgan, Nigel, and Linda Davies, "How Cephalonia Can Conquer the World (Or at the Very Least, Your Students!): A Library Orientation Case Study from Cardiff University." In Douglas Cook and Ryan L. Sittler (Eds.) *Practical Pedagogy for Library Instructors: 17 Innovative Strategies to Improve Student Learning*. Chicago, IL: Association of College and Research Libraries, 2008.

National Institutes of Health (NIH). "NOT-OD-10-019: Update on the Requirement for Instruction in the Responsible Conduct of Research." Published November 24, 2009. https://grants.nih.gov/grants/guide/notice-files/NOT-OD-10-019.html.

National Library of Medicine (NLM). "OLDMEDLINE Data." Published December 29, 2017. https://www.nlm.nih.gov/databases/databases_oldmedline.html.

NIH Office of the Director. "NIGMS Clearinghouse for Training Modules to Enhance Data Reproducibility: NIH Rigor and Reproducibility Training Modules." Published 2015. https://www.nigms.nih.gov/training/pages/clearinghouse-for-training-modules-to-enhance-data-reproducibility.aspx#nih-rigor-and-reproducibility-training-modules.

Office for Human Research Protections (OHRP), Department of Health and Human Services (DHHS). "Guidance on Continuing Review – January 2007," HHS.gov. Published April 7, 2016. https://www.hhs.gov/ohrp/regulations-and-policy/guidance/continuing-review-january-2007/index.html.

Phillips, Trisha, Franchesca Nestor, Gillian Beach, and Elizabeth Heitman. "America COMPETES at 5 Years: An Analysis of Research-Intensive Universities' RCR Training Plans." *Science and Engineering Ethics* 24, no. 1 (2018): 227–49, https://doi.org/10.1007/s11948-017-9883-5.

Read, Kevin, and Alisa Surkis. "Research Data Management Teaching Toolkit." FigShare.com. Posted February 1, 2018. https://doi.org/10.6084/m9.figshare.5042998.v6.

Resnik, David B., and Gregg E. Dinse, "Do U.S. Research Institutions Meet or Exceed Federal Mandates for Instruction in Responsible Conduct of Research? A National Survey," *Academic Medicine: Journal of the Association of American Medical Colleges* 87, no. 9 (September 2012): 1237–42, https://doi.org/10.1097/ACM.0b013e318260fe5c.

Rosnow, Ralph L. "Teaching Research Ethics through Role-Play and Discussion," *Teaching of Psychology* 17, no. 3(1990): 179–181. https://doi.org/10.1207/s15328023top1703_10.

Steward, Oswald, and Rita Balice-Gordon. "Rigor or Mortis: Best Practices for Pre-clinical Research in Neuroscience." *Neuron* 84, no. 3 (2014): 572–81, https://doi.org/10.1016/j.neuron.2014.10.042.

White, Marijane, and Jean-Philippe Gourdine, "Data Rigor and Reproducibility" (Course materials for OHSU PMBC PERT Data Rigor and Reproducibility nano-course). Published July 28, 2020. https://doi.org/10.17605/OSF.IO/TEGY9.

Yoon, Ayoung, and Teresa Schultz. "Research Data Management Services in Academic Libraries in the US: A Content Analysis of Libraries' Websites." *College & Research Libraries* 78, no. 7(2017): 920. https://doi.org/10.5860/crl.78.7.920.

Chapter Nine

Animal Care and Use in Veterinary Teaching and Clinical Research

Faythe Thurman and Marcelle Savoy

Veterinary medical education and veterinary clinical research present unique challenges and considerations to those who are members of Institutional Animal Care and Use Committees (IACUCs) and those who support faculty or researchers creating IACUC protocols. To produce day-one ready veterinarians, the education of these students necessitates learning skills on live animals. Animal use in training veterinarians has varied throughout the years and is still a topic of discussion in the field. The 2020 Association of American Veterinary Medical Conference and Iverson Bell Symposium included "The Use of Animals in Education Symposium," a forum to discuss and evaluate how animals, models, and simulations are used in the training of veterinarians.[1] This important conversation in the field of veterinary education addresses the delicate balance between the number of procedures performed to produce skilled veterinarians and animal well-being. Some institutions work with limited herds, which makes for a challenging task of scheduling students requiring clinical procedures training to correlate with rest periods for the animals.

In addition to teaching, if schools and colleges of veterinary medicine are connected to Level 6 universities (having doctoral programs that require dissertations), they must participate in compulsory research.[2] Veterinary medical research creates opportunities to improve animal health and welfare; understand disease etiology and progression; and gain knowledge leading to new methodologies, techniques, and treatment, providing a potential path for translation into human medicine. As with teaching programs, having enough animal subjects to meet the 3Rs directive (replacement, reduction, and refinement; as proposed by Russell and Burch in 1959) is challenging and requires unique approaches to humane research experimentation. This chapter outlines different approaches by veterinary teaching colleges, which emphasize applied clinical research, to deliver effective student training and produce high-quality scholarly output while under the auspices of the IACUC.

VETERINARY MEDICAL EDUCATION AND THE IACUC

History

The seeds of modern veterinary education in the United States were planted shortly after the Civil War in response to the cruel treatment of horses and mules.[3] In this account, Willems also documented the founding of the Massachusetts Society for the Prevention of Cruelty to Animals in 1868 by George Thorndike Angell, which is now known as America's first humane organization. In the 1880s, veterinary education consisted of a total of twelve non-consecutive months of study over a two-year period with the ability to read and write as the sole prerequisite, but by the late 1940s it evolved into two years of pre-veterinary coursework and four years of professional academics, a standard that is held today.[4]

Nearly one hundred years after the foundation of country's first humane organization, the Animal Welfare Act (AWA) was enacted by Congress under President Johnson in 1966 as the regulatory agency for animal use in colleges and university, including veterinary schools. Initially, the AWA was enacted to prohibit the illegal trafficking of stolen pets to research laboratories, but because of an added amendment in 1970, the agency evolved into regulatory oversight for animal care and use during experimentation and teaching.[5] Another amendment in 1985 brought forth the formation of the IACUC.

For animal care use protocols (ACUPs), methods to mitigate pain and distress through implementation of the 3Rs required robust searching and documentation through the known literature.[6] Therefore, it is not uncommon to have librarians as active members on the IACUC.

From 1986 through most of the 1990s, techniques in pain management via applications of the 3Rs for animal use were still in its infancy but beginning to emerge into well-defined standards.[7] However, pain-producing procedures and distress from excessive animal handling[8] that might be incurred during teaching were not sufficiently outlined through the regulatory process and had been highly debated among veterinary colleges.[9] Searching the literature for pain management has been particularly challenging for librarians who sit on IACU committees and who serve as liaisons for colleges of veterinary medicine. Even as late as 2018, federal laws and policies regarding pain management still target research facilities and lack specificity toward addressing

ACUPs for teaching purposes (§2.36.1-8, 9 CFR AWR (1-1-18 edition)) indicating that AWA regulations are at most parenthetical.[10]

Veterinary Education IACUC Protocols

Protocols to cover the use of live animals in labs to teach clinical skills are brought to the IACUC annually for either new approval or annual review. These protocols create many challenges for the primary investigators (PIs) writing them as well as the IACUCs reviewing them. The greatest difficulty PIs face is that the questions asked in the protocol forms are focused on research uses for animals not teaching.[11] Several common stumbling blocks include over-generalized rationales, arbitrary justifications for numbers of animals used, and inadequate literature searches with little discussion of the literature found. This section will discuss what to look for in a strong veterinary teaching protocol and recommendations on ways to help PIs improve these areas.

The overview and rationale sections of an IACUC protocol should provide a strong narrative description and reasoning for the proposed use of the animals in lay terms. Often in veterinary teaching protocols, this narrative is weak when it is a brief explanation that boils down to "students need to learn these skills on real animals." This reason is a logical one, but not robust enough. Creating such a narrative is challenging given that these are often procedures done in labs that are required in the curriculum and by accreditors with not much more immediate reasoning. PIs should be encouraged to discuss ways that they have reduced animal use, via simulation models and other alternative teaching methods, that make the live animal the final step in solidifying their learning experience and accomplish the goal of creating veterinarians that are ready to practice with real-life experience under their belts. Another helpful addition to this part of a protocol is to have PIs do a cost-benefit analysis of sorts, answering questions like What are the risks to the animals? What are the benefits that the animals, students, and society get? Is the risk worth the benefit?

In a similar way, the justification of numbers of animals to be used is often a difficult narrative to construct. Some scenarios give PIs clear calculations and reasons such as a limited number of animals in a herd that is far less than the number of students in a lab. A veterinary college might have a bovine teaching herd of twenty-five cattle and fifty students resulting in a ratio of two students to one cow; but what would this ratio look like for the same number of students using four hundred—plus local, privately owned cattle temporarily provided by the proprietors to the veterinary college for educational purposes? A one-to-one ratio is the perfect scenario, but should twenty-five more cattle be involved when other veterinary colleges teach with less? The real challenge in these situations is that no clear guidelines or research exists detailing the ideal ratios of animals to students in a veterinary teaching environment. Pedagogic research in veterinary teaching laboratories is limited, especially in the areas of student number accommodations and the effect of certain procedures on different animal species, to help determine the optimal animal to student ratio.[12] Until the research becomes more definitive, PIs should provide justifications that include animal population limits, spacing limitations, safe handling as a team (multiple students per animal to handle it while a student performs a procedure), general desire to reduce animal use, and other relevant restrictions on the use of animals.

The literature search required in an IACUC protocol is the area where librarian members, whether or not they serve on the committee, have the most to contribute. They can provide support in the search evaluation process to faculty members who are required to submit these protocols as part of their academic obligation. The main goal of the literature search is to generate a thorough review of the existing literature so the PI understands what alternatives may exist for the animals and procedures that will be conducted. In IACUC protocols, the sections for the literature search often use two different phrases in the title and/or questions asked: "reduction, refinement, and replacement" (the 3Rs) or "alternatives." At Lincoln Memorial University, for example, we have two sections pertaining to the literature search. They are titled "Literature review for refinement, reduction, and replacement of the use of animals" and "Refinement, reduction, and replacement of the use of animals." The first section asks the PI to provide information on what resources were searched (databases and websites, though a minimum of two databases is required), the date of the search, what search strings were used, and the number of results per search. The second section is a place for PIs to discuss what they found in the literature review in regard to possible alternatives that would reduce the number of animals; refine the procedures, techniques, and materials used in order to reduce animal pain, distress, or discomfort; or replace one species with another.

These searches can be in an area where PIs for veterinary education protocols do not expend a lot of effort as they do not see the relevance in education when student learning and accreditation may require the use of live animals at some point for students to learn skills. In their mind, no replacements or alternatives exist to the live animals at the point where students must apply what they have learned through lecture, demonstration, practice on models, or other alternative teaching methods to a real animal to demonstrate competency and gain experience with a patient that will move, possible resist, and require the student to use their animal handling knowledge to complete the task.

As these literature searches pertain to the 3Rs and alternatives, many PIs tend to the term "alternative" in their search thinking that it is the best way to get to the literature they need to review. In veterinary education protocols, a poor alternatives search might look like "veterinary education alternatives." This method of searching would not meet the requirements of the Animal Welfare Act. In fact, the term "alternative" should be avoided in most searches, especially veterinary education protocols. The U.S. Department of Agriculture's (USDA) Animal Welfare Information Center (AWIC) has compiled the Animal Use Alternative Thesaurus to aid in finding appropriate terms for 3Rs searches for IACUC protocols.[13] This thesaurus includes terms that would be particularly useful for veterinary education such as "animal models," "simulation," and "virtual reality" as general animal use alternatives. "Blood collection," "surgery," and "anesthesia" are examples of terms for procedures or techniques that could be part of labs.

For a protocol covering teaching with an equine herd, a robust and appropriate search might look like "(education OR training) AND (equine OR horse) AND (simulation OR model OR software)." For a specific procedure such as equine ultrasounds, "(education OR training) AND (equine OR horse) AND (ultrasound OR imaging) AND (simulation OR model OR software OR "virtual reality")." PIs may want to take out the species-specific terms to see if there are models or simulations from other species, including humans, that might be translated to the species they are working with. If they would like to keep the search in the animal realm, they should consider replacing the specific species with the term "veterinary." In general, this direction of searching does mean that multiple searches may need to be conducted and documented, but this technique of searching shows an IACUC and inspectors that PIs are serious and thorough in their exploration of the literature.

The good news is that often once a good search strategy for a veterinary education protocol is established, the process becomes simpler and easier in the future. A strong base means that the search can simply be replicated every year and the newest sources reviewed for any innovative teaching methods, pain management, procedures, and so forth. Periodically, the PI may want to review their search with a librarian to see if it can be improved or updated based on developments in the field. Changes or additions to the curriculum or teaching environment are another logical point to completely review and update search strategies.

While PIs and IACUCs face difficulties in constructing and evaluating animal care and use protocols for veterinary education, the most important factor in overcoming these challenges is communication. Expectations from the IACUC concerning these protocols develop over time and should be communicated clearly to PIs with specialized training or guidance as needed.

IACUCs may also want to consider separate protocol forms for veterinary education where most of the content is the same as research-focused protocol forms but select questions have been modified to provide clarity on the information needed by the IACUC in regards to animal care and use in veterinary education.

Resources for Searching Alternatives to Support the 3Rs for Teaching

Combing through the literature of peer-reviewed journal articles in mostly standard veterinary databases (e.g., CAB Abstracts with CABI, AGRICOLA, and PubMed) for alternatives is an essential process for IACUC protocols. Because the application of the 3Rs to veterinary education is not as prolific in the literature as it is in animal research, following the model of scoping reviews and extending the bounds of the search to atypical sources may lead to better results. Examples include the following, which are discussed individually.

1. University Websites
    The consumptive use of animals in laboratory training has come under scrutiny of regulatory agencies, animal rights groups, and veterinary students and in response, veterinary educators developed novel alternative teaching techniques to address this evolving issue.[14] Within their commentary, Hart and co-authors compiled a list of alternative teaching procedures contributed by veterinary educators.[15] Information on these alternatives can be found on the websites of the contributing universities and colleges.
    One such resource of interest is located at the Center for Animal Resources and Education (CARE) website at Cornell University. The Translational Training Tools™ developed there are presented in two volumes of "recipes" for crafting training tools used for non-surgical and surgical procedures.[16] Even when similar tools are created in-house by an institution and implemented into teaching and training, insight into procedures used by other institutions can be valuable to instructors.
    Simulations may offer another approach to alternative use of animals adopted by various schools and colleges of veterinary medicine. Medical simulations are scenarios that aim to model real life situations and are classified along a spectrum from high- to low-fidelity (most to least accurate, respectively).[17] Washington State University provides an exemplar website dedicated to the application of simulations into their veterinary medicine curriculum.[18] To reduce the usage of living animal subjects, librarians can direct veterinary educators to similar sites for the purpose

of supporting the 3Rs by integrating simulation into student training. The WSU Clinical Simulation Center website also contains a wish list page for other desirable simulators that may be funded through a school's annual budget or from a donations program. The value of having such a resource list may prove indispensable as veterinary educators continue to seek diverse avenues for reducing, replacing, and refining the use of live animals in their teaching programs.

2. Federal Agencies

   Handling and restraint is an essential component of veterinary education, but often leads to stress, pain, agitation, and aggression in animals if performed excessively.[19] Federal code of regulations Handling of Animals 9 CFR AWR § 2.131[20] supply regulations but with scant guidelines or instructions for best practices. Although instructors are effective in this area, a productive search using unconventional sources may result in serendipitous findings of new techniques satisfying the 3Rs requirement in teaching protocols. Take for example the role of the extension veterinarian specialist. The Cooperative Extension System began in 1914 with the Smith-Lever Act, a partnership between the USDA and each state's land-grant university, whose mission includes the teaching of agriculture.[21] Extension veterinarians provide outreach and continuing education on best practices and research to producers and practitioners. Extension bulletins are usually written in layman's or non-technical language and are usually posted on the department's website or on a Lib-guide created by the liaison librarian. A bulletin from Penn State Extension, for example, details handling procedures for food animals[22] and although the plain language targets producers, the novice veterinary student will easily benefit from this simplicity.

3. Gray Literature

   Unpublished or non-commercially published research is defined as gray literature and comprises mostly high-quality up-to-date studies configured as government reports, conference proceedings, theses, dissertations and other non-conventional forms.[23] Indexes of grey literature are located at specific sites such as CORE, Bielefeld Academic Search Engine (BASE), and Open Grey. Many of these databases also link to international institutional repositories (IRs) that contain a plethora of grey literature. For example, a narrow search in the CORE database filtered by title [*title:((animal AND alternatives AND teaching))*] retrieved three relevant records from (a) the IR located at the University of Sydney, (b) IntechOpen book repository, and (c) CiteSeerX public digital library.

In summary, thinking outside the box of standard database searching may help educators break through the impasse of non-substantive retrievals as they continue their search for ways to incorporate the 3Rs into their teaching methodologies.

VETERINARY CLINICAL RESEARCH

Veterinary clinical research is varied in its methods, scope, animal species, and applications. Unlike larger clinical trials found in the Veterinary Information Network (VIN) Clinical Trials Database, clinical studies at Lincoln Memorial University College of Veterinary Medicine are conducted on a significantly smaller scale. Effectiveness of nutritional supplements for various ailments, therapy interventions, and surgical procedures are only a few of the research topics that can be addressed in small-scale clinical investigations.[24] But regardless of the sampling size, the benefits from well-designed clinical studies help propel veterinary medical students toward evidence-based decision making in veterinary practice.[25]

Unlike teaching protocols where regulations are vague and tangential, clinical research studies are regulated more stringently by the Public Health Service (PHS), the AWA, and the Federal Food, Drug, and Cosmetic Act.[26] Research protocols may amplify the librarian's role from assessing proper documentation of alternatives to assisting with the different components of the clinical research itself, including the literature search, location of grant-funding organizations, compliance with ethical and legislative principles, data management, and writing style.[27]

For the IACUC, some important distinctions can be made between institution-owned animals and client-owned animals that may impact protocol submission at veterinary schools. All institution-owned animals, as defined by the Animal Welfare Act, require an IACUC protocol for teaching and research purposes. Mandatory submission of an IACUC protocol varies with client-owned animals depending on the nature of the research. Generally, if the research is being conducted as a part of the patient's standard care, then a protocol is not required as the patient is receiving clinical care they would have received anyways during their visit. An example is use of tissue or other medical samples after procedures that were part of a patient's diagnostic and treatment plan. If dogs have surgery to remove tumors and research will be done with the tumors after they are taken from the dogs, the researcher does not need to submit a protocol as long as the removal of the tumors was a part of normal treatment for the health of the patient. If, for example, the research is designed to ask clients to allow blood samples to be taken from their animals whether or not the sampling is medically necessary, this research is more experimental and would require an IACUC protocol.[28]

Even if the research does not require IACUC review, all veterinary clinical researchers still should be encouraged to submit IACUC protocols. Some journals that the investigator may want to submit their research to may require an approved IACUC protocol, but protocols can only be approved prior to the activities discussed in the research as the animal welfare must be monitored throughout the process. No retroactive approvals should be granted by an IACUC. When discussing IACUC protocols with researchers, a clear benefit of submitting a protocol is the opening of as many publishing options with as few obstacles as possible.

CONCLUSION

Treating animals humanely in American veterinary education had its humble beginnings after the Civil War in response to animal cruelty. As medical science progressed through the years so did the use of inhumane animal experimentation prompting the development of the 3Rs by Russell and Burch in 1959 in Great Britain, followed by the 1966 Animal Welfare Act legislation in the United States. Although rigid guidelines have been established for animal research, the parameters have not been as clear in the area of teaching and training student veterinarians. There has been much disagreement among veterinary schools concerning the use of animals in teaching, especially in the area of reduction. Russell and Burch acknowledged that the 3Rs may not always be in balance by advocating at times that refinement should take precedence over reduction, especially if the latter is applied to pain and distress and not necessarily animal numbers.[29] Even within this ambiguity, librarians have the obligation to perform relevant searches in traditional and innovative venues for best practices in veterinary education as they continue to serve and address the teaching and research protocols set forth by the Institutional Animal Care and Use Committee.

NOTES

1. Association of American Veterinary Medical Colleges, *2020 Annual Conference Summary*, 2020.

2. Marla Stephens, Laura K. Warren, and Ariana L. Harner, *Comparative Indicators of Education in the United States and Other G-20 Countries: 2015* (NCES 2016-100) (National Center for Education Statistics [NCES], December 2015).

3. Robert A. Willems, "Animals in Veterinary Medical Teaching: Compliance and Regulatory Issues, the US Perspective," *Journal of Veterinary Medical Education* 34, no. 5 (Winter 2007): 615–19, https://doi.org/10.3138/jvme.34.5.615.

4. Oscar J. Fletcher, Billy E. Hooper, and Regina Schoenfeld-Tacher, "Instruction and Curriculum in Veterinary Medical Education: A 50-Year Perspective," *Journal of Veterinary Medical Education* 42, no. 5 (Winter 2015): 489–500, https://doi .org/10.3138/jvme.0515-071.

5. Lynn C. Anderson, "Institutional and IACUC Responsibilities for Animal Care and Use Education and Training Programs," *ILAR Journal* 48, no. 2 (2007): 90–95, https://doi.org/10.1093/ilar.48.2.90.

6. Daureen Nesdill and Kristina M. Adams, "Literature Search Strategies to Comply with Institutional Animal Care and Use Committee Review Requirements," *Journal of Veterinary Medical Education* 38, no. 2 (Summer 2011): 150–56, https:// doi.org/10.3138/jvme.38.2.150.

7. Dale F. Schwindaman, "The History of the Animal Welfare Act," 147–51, in *50 Years of Laboratory Animal Science,* published by the American Association for Laboratory Animal Science (Memphis, TN: American Association for Laboratory Animal Science, 1999).

8. Kevin J. Stafford and Vicki H. Erceg, "Teaching Animal Handling to Veterinary Students at Massey University, New Zealand," *Journal of Veterinary Medical Education* 34, no. 5 (Winter 2007): 583–85, https://doi.org/10.3138/jvme.34.5.583.

9. Stuart E. Leland, Pamela A. Straeter, and Beverly Jan Gnadt, "The Role of the IACUC in the Absence of Regulatory Guidance," *ILAR Journal* 60, no. 1 (2019): 95–104, https://doi.org/10.1093/ilar/ilz003.

10. Tracy H. Vermulapalli, Shawn S. Donkin, Timoty B. Lescun, Peggy A. O'Neil, and Patrick A. Zollner, "Considerations When Writing and Reviewing a Higher Education Teaching Protocol Involving Animals," *Journal of the American Association for Laboratory Animal Science* 56, no. 5 (2017): 500–508.

11. Ibid.

12. Ibid.

13. Animal Welfare Information Center (AWIC), *Animal Use Alternative Thesaurus Terminology - Alphabetical Listing,* 2015, https://pubs.nal.usda.gov/animal-use -alternatives-thesaurus-terminology-alphabetical-listing.

14. Lynette A. Hart, Mary W. Wood, and Hsin-Yi Weng, "Mainstreaming Alternatives in Veterinary Medical Education: Resource Development and Curricular Reform," *Journal of Veterinary Medical Education* 32, no. 4 (Winter 2005): 473–80, https://doi.org/10.3138/jvme.32.4.473.

15. Ibid.

16. Wendy O. Williams, David E. Mooneyhan, and Christine M. Peterson, *Translational Training Tools: The 3 Ts Serving the 3 Rs: Volume 1: Recipes for Crafting Your Own Purpose-Specific Training Tools for Non-Surgical Procedures* (Center for Animal Resources and Education at Cornell University, 2015a), https://ras.research. cornell.edu/care/3T.html; Wendy O. Williams, David E. Mooneyhan, and Christine M. Peterson, *Translational Training Tools: The 3 Ts Serving the 3 Rs: Volume 1: Recipes for Crafting Your Own Purpose-Specific Training Tools for Surgery Practice* (Center for Animal Resources and Education at Cornell University, 2015b), https:// ras.research.cornell.edu/care/3T.html.

17. Ross J. Scalese and S. Barry Issenberg, "Effective Use of Simulations for the Teaching and Acquisition of Veterinary Professional and Clinical Skills," *Journal of Veterinary Medical Education* 32, no. 4 (Winter 2005): 461–67, https://doi.org/10.3138/jvme.32.4.461.

18. Washington State University, "WSU College of Veterinary Medicine Clinical Simulation Center," https://www.vetmed.wsu.edu/innovative-education/clinical-simulation-center.

19. Stafford and Erceg, "Teaching Animal Handling to Veterinary Students at Massey University, New Zealand."

20. United States Department of Agriculture Animal and Plant Health Inspection Service, *USDA Animal Care: Animal Welfare Act and Animal Welfare Regulations* (Washington, DC: United States, Department of Agriculture, Animal and Plant Health Inspection, 2017), https://www.aphis.usda.gov/animal_welfare/downloads/bluebook-ac-awa.pdf.

21. Katie Burns, "Rumors of the Demise of the Extension Veterinarian," *JAVMA News* (November 15, 2015), https://www.avma.org/javma-news/2015-11-15/rumors-demise-extension-veterinarian.

22. Dennis Murphy, "Animal Handling Tips," Penn State Extension, last updated June 12, 2014, https://extension.psu.edu/animal-handling-tips#nav-12.

23. University of New England, "Grey Literature," 2021, https://www.une.edu.au/library/support/eskills-plus/research-skills/grey-literature.

24. American Veterinary Medical Association (AVMA), "Find Veterinary Clinical Trials," https://www.avma.org/resources-tools/find-veterinary-clinical-studies, 2021; Ping Deng and Kelly S. Swanson, "Companion Animals Symposium: Future Aspects and Perceptions of Companion Animal Nutrition and Sustainability," *Journal of Animal Science* 93, no. 3 (2015): 823–34, https://doi.org/10.2527/jas.2014-8520.

25. Frances Barr, "Masters in Clinical Veterinary Research," *Veterinary Record* 179, no. 8 (2016): i–ii, https://doi.org/10.1136/vr.i4522.

26. Victoria A. Hampshire, "Regulatory Issues Surrounding the Use of Companion Animals in Clinical Investigations, Trials, and Studies," *ILAR Journal* 44, no. 3 (2003): 191–96, https://doi.org/10.1093/ilar.44.3.191.

27. Barr, "Masters in Clinical Veterinary Research."

28. Patricia Brown and Chester Gibson, "Response to Protocol Review Scenario: A Word from OLAW and USDA," *Lab Animal* 39, no. 3 (March 2010): 3, https://www.nature.com/articles/laban0310-68b.

29. Jerrold Tannenbaum and B. Taylor Bennett, "Russell and Burch's 3Rs Then and Now: The Need for Clarity in Definition and Purpose," *Journal of the American Association for Laboratory Animal Science* 54, no. 2 (2015): 120–32, https://www.ncbi.nlm.nih.gov/pmc/articles/PMC4382615/.

BIBLIOGRAPHY

American Veterinary Medical Association (AVMA). "Find Veterinary Clinical Trials." 2021. https://www.avma.org/resources-tools/find-veterinary-clinical-studies.

Anderson, Lynn C. "Institutional and IACUC Responsibilities for Animal Care and Use Education and Training Programs." *ILAR Journal* 48, no. 2 (2007): 90–95. https://doi.org/10.1093/ilar.48.2.90.

Animal Welfare Information Center (AWIC). *Animal Use Alternative Thesaurus Terminology - Alphabetical Listing.* Published 2015. https://pubs.nal.usda.gov/animal-use-alternatives-thesaurus-terminology-alphabetical-listing.

Association of American Veterinary Medical Colleges. *2020 Annual Conference Summary.* Published 2020. https://www.aavmc.org/meetings/2020-annual-conference-summary.

Barr, Frances. "Masters in Clinical Veterinary Research." *Veterinary Record* 179, no. 8 (2016): i–ii. https://doi.org/10.1136/vr.i4522.

Brown, Patricia, and Chester Gibson. "Response to Protocol Review Scenario: A Word from OLAW and USDA." *Lab Animal* 39, no. 3 (March 2010): 3. https://www.nature.com/articles/laban0310-68b.

Burns, Katie. "Rumors of the Demise of the Extension Veterinarian." *JAVMA News* (November 15, 2015). https://www.avma.org/javma-news/2015-11-15/rumors-demise-extension-veterinarian.

Fletcher, Oscar J., Billy E. Hooper, and Regina Schoenfeld-Tacher. "Instruction and Curriculum in Veterinary Medical Education: A 50-Year Perspective." *Journal of Veterinary Medical Education* 42, no. 5 (Winter 2015): 489–500. https://doi.org/10.3138/jvme.0515-071.

Hampshire, Victoria A. "Regulatory Issues Surrounding the Use of Companion Animals in Clinical Investigations, Trials, and Studies." *ILAR Journal* 44, no. 3(2993): 191–96. https://doi.org/10.1093/ilar.44.3.191.

Hart, Lynette A., Mary W. Wood, and Hsin-Yi Wang. "Mainstreaming Alternatives in Veterinary Medical Education: Resource Development and Curricular Reform." *Journal of Veterinary Medical Education* 32, no. 4 (Winter 2005): 473–80. https://doi.org/10.3138/jvme.32.4.473.

Leland, Stuart E., Pamela A. Straeter, and Beverly Jan Gnadt, "The Role of the IACUC in the Absence of Regulatory Guidance." *ILAR Journal* 60, no. 1 (2019): 95–104. https://doi.org/10.1093/ilar/ilz003.

Murphy, Dennis. "Animal Handling Tips." Penn State Extension, last updated June 12, 2014. https://extension.psu.edu/animal-handling-tips#nav-12.

Nesdill, Daureen, and Kristina M. Adams. "Literature Search Strategies to Comply with Institutional Animal Care and Use Committee Review Requirements." *Journal of Veterinary Medical Education* 38, no. 2 (Summer 2011): 150–56. https://doi.org/10.3138/jvme.38.2.150.

Scalese, Ross J., and S. Barry Issenberg. "Effective Use of Simulations for the Teaching and Acquisition of Veterinary Professional and Clinical Skills." *Journal of Veterinary Medical Education* 32, no. 4 (Winter 2005): 461–67. https://doi.org/10.3138/jvme.32.4.461.

Schwindaman, Dale F. "The History of the Animal Welfare Act" (147–151). In American Association for Laboratory Animal Science, *50 Years of Laboratory Animal Science* (Memphis, TN: American Association for Laboratory Animal Science, 1999).

Stafford, Kevin J., and Vicki H. Erceg. "Teaching Animal Handling to Veterinary Students at Massey University, New Zealand." *Journal of Veterinary Medical Education* 34, no. 5 (Winter 2007): 583–85. https://doi.org/10.3138/jvme.34.5.583.

Stephens, Marla, Laura K. Warren, and Ariana L. Harner. *Comparative Indicators of Education in the United States and Other G-20 Countries: 2015.* National Center for Education Statistics (NCES), December 2015.

Tannenbaum, Jerrold, and B. Taylor Bennett. "Russell and Burch's 3Rs Then and Now: The Need for Clarity in Definition and Purpose." *Journal of the American Association for Laboratory Animal Science* 54, no. 2 (2015): 120–32. https://www.ncbi.nlm.nih.gov/pmc/articles/PMC4382615/.

United States Department of Agriculture Animal and Plant Health Inspection Service. *USDA Animal Care: Animal Welfare Act and Animal Welfare Regulations.* Washington, DC: United States, Department of Agriculture, Animal and Plant Health Inspection Service, 2017. https://www.aphis.usda.gov/animal_welfare/downloads/bluebook-ac-awa.pdf.

University of New England. "Grey Literature." 2015. https://www.une.edu.au/library/support/eskills-plus/research-skills/grey-literature.

Vemulapalli, Tracy H., Shawn S. Donkin, Timothy B. Lescun, Peggy A. O'Neil, and Patrick A. Zollner. "Considerations When Writing and Reviewing a Higher Education Teaching Protocol Involving Animals." *Journal of the American Association for Laboratory Animal Science* 56, no. 5 (2017): 500–508.

Washington State University. "WSU College of Veterinary Medicine Clinical Simulation Center." https://www.vetmed.wsu.edu/innovative-education/clinical-simulation-center.

Willems, Robert A. "Animals in Veterinary Medical Teaching: Compliance and Regulatory Issues, the US Perspective." *Journal of Veterinary Medical Education* 34, no. 5 (Winter 2007): 615–19. https://doi.org/10.3138/jvme.34.5.615.

Williams, Wendy O., David E. Mooneyhan, and Christine M. Peterson. *Translational Training Tools: The 3 Ts Serving the 3 Rs: Volume 1: Recipes for Crafting Your Own Purpose-Specific Training Tools for Non-Surgical Procedures.* Center for Animal Resources and Education at Cornell University. 2015a. https://ras.research.cornell.edu/care/3T.html.

Williams, Wendy O., David E. Mooneyhan, and Christine M. Peterson. *Translational Training Tools: The 3 Ts Serving the 3 Rs: Volume 1: Recipes for Crafting Your Own Purpose-Specific Training Tools for Surgery Practice.* Center for Animal Resources and Education at Cornell University. 2015b. https://ras.research.cornell.edu/care/3T.html.

Chapter Ten

An Informed Consent Document Review Service

Michele Nance, Emily F. Gorman, and Everly D. Brown

Institutional review boards (IRBs) arose out of a need to protect the rights and welfare of people participating in research following years of unethical human research experiments such as the Tuskegee syphilis study.[1] A major component of these protections is to ensure participants understand the purpose of the research being done, the procedures involved, the risks to the participants, what will be done with any data and/or biological samples that are collected, and the fact that participation is voluntary and can be stopped at any time without penalty. The process of communicating this information to potential research participants is known as informed consent.[2] Informed consent typically involves the use of written documents that participants read and then sign to acknowledge their agreement to be in the study.[3]

While it sounds straightforward, composing an effective informed consent form (ICF) can be difficult. The study team must balance the need to communicate all the pertinent information about the research with the need to make this information comprehensible to all participants. The federal requirements do not mandate a specific reading level for ICFs; they only require that informed consent be "in language understandable to the subject."[4] The average reading level of adults in the United States is equivalent to an eighth-grade level.[5] As a result, most institutions recommend that ICFs be written at this level or below; however, that is easier said than done.[6] In reality, numerous studies have shown that ICFs do not meet this readability standard, particularly for health science research.[7] One study found that the readability of cancer research ICFs has actually decreased over time.[8]

One likely reason many consent forms have a high reading level is that they are usually written by the study team, who are so immersed in their topic of study that they may not realize they are using too much jargon or referring to a concept or procedure that may be unfamiliar to the average person.

For medical studies in particular, the procedures and risks are often riddled with professional terminology that nonscientists would not know.[9] One way study teams can avoid the jargon pitfall is by having their ICFs reviewed by a layperson, someone who can identify potentially unfamiliar procedures and terms that could benefit from definition, editing, or otherwise rephrasing. Libraries have the opportunity to step into this review role for researchers at their institutions. This chapter describes how the Health Sciences and Human Services Library (HSHSL) at the University of Maryland, Baltimore (UMB) collaborated with their Human Research Protections Office (HRPO) to implement an ICF review service.

BACKGROUND ON THE HSHSL'S SERVICE

When the consent form review service was introduced in 2011, the HSHSL had two public service desks: one for reference and research services, and the other for circulation services. At that time, many academic health sciences libraries were moving away from this traditional two-desk service model in favor of a merged-desk model, in which reference and circulation services are offered at a single service point. This trend was driven by a gradual yet notable decrease in the number of both reference and circulation transactions. As more library resources became available online, as students and faculty gained more facility with online research, and as technology within the library became more reliable, library users developed a greater degree of self-sufficiency. As a result, many academic health sciences libraries at the time—the HSHSL included—reported annual declines in the number of reference questions answered and items circulated and re-shelved.

In 2011, in the wake of the 2008 financial crisis, the university was continuing to implement 2-percent pay cuts across the board for all university staff, in response to ongoing budget cuts throughout the University System of Maryland. Nevertheless, it became evident the salary reductions would not be sufficient and additional measures would be necessary. The HSHSL, like other campus entities, came under review by campus administration and was asked to demonstrate how library services added value by supporting the university's education and research mission. It was in this environment, with reduced traffic at the reference desk and an imperative to contribute a broader range of services with greater efficiency, that the head of reference services at HSHSL proposed a new service: reviewing consent forms for campus researchers and editing them to improve readability.

The HSHSL created an online request form[10] for researchers to submit consent documents for review. This form asks requesters to give their contact information, affiliation, and a need-by date. They can also add comments and attach up to three documents for review. The service was promoted to the campus and to the HRPO. The HRPO was impressed enough with the service to highlight it on their website for researchers, stating, "As a free service to all UMB faculty, researchers are strongly encouraged to use this service."[11] As forms are submitted, a select group of three trained staff members in the Information Services department receive an email with the consent documents attached. In the beginning, one staff member would claim a request by emailing the rest of the group. Staff reviewed the document using Microsoft Word's proofing option, which allows users to view readability statistics using the Flesch-Kincaid reading ease and grade-level tests. They also used Word's track changes feature to make edits and offer suggestions.

In 2018, staff decided to update the ICF review submission system to collect more detailed statistics and to monitor the consent form process from start to finish. By March 2019, the library's information technology department developed an online portal that gave each incoming request a unique ID and initiated an email to staff. When a staff member enters the portal and claims a new job, the system notifies the team and moves the claimed job from a "New" to an "In Progress" status. The request form contains a number of required fields to complete, which serve to enrich our statistics. These include requester status, title of study, length of documents, number of documents, pre-review reading level, post-review reading level, time spent editing (in minutes), and date returned. After the edited consent form is returned to the requester, the job is moved into a "Completed" status. After seven days, the system sends a survey to the requester to inquire if the edits were useful and if the requester would recommend the service to others. There has been a 100 percent positive response to the survey thus far, and respondent feedback has include such comments as "The edits were all incorporated into the final consent forms we submitted to the IRB," "Excellent," and "Very nice people with good ideas."

In the ten years that the HSHSL has offered the consent form review service, there have never been more than three library reviewers at any one time, and for a ten-month period in 2017–2018, the review team was down two members. Consent form review requests continue to grow as the years progress, increasing from sixteen in 2012 (the first full year of the service) to forty-six in 2020 (see figure 10.1). The average number of pages in each submission also increased from 6.0 in 2012 to 10.6 in 2020. The longest ICF submitted to date came in at forty-five pages. The team began tracking the time we spent working on these documents in 2018. In 2020, the last complete year of data collected on the service, we spent a total of 316 hours reviewing consent forms, an average of 6.87 hours per form.

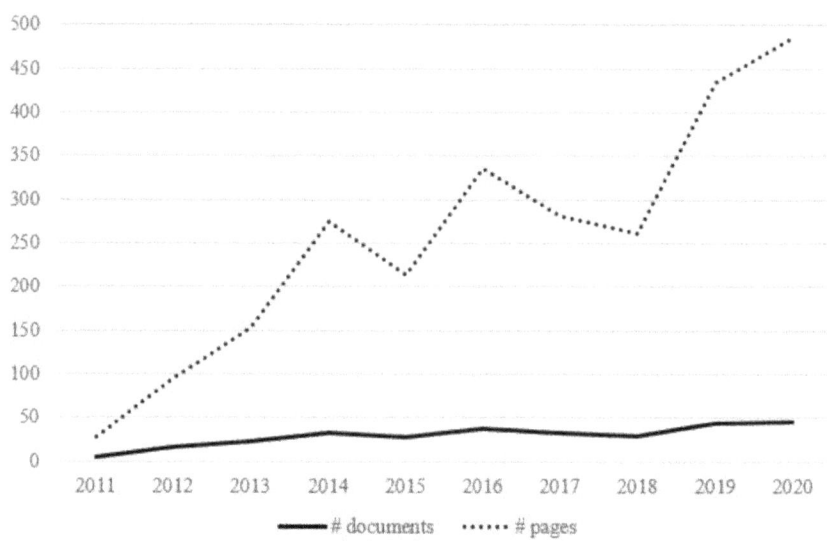

Figure 10.1. Number of Documents and Pages Reviewed per Year.
Michele Nance, Emily F. Gorman, and Everly D. Brown

CHALLENGES OF THE SERVICE

Time, Schedules, and Staffing

When the service began in 2011, it was offered by staff working at the library's reference desk. The reference desk was open from 8:00 a.m. to 8:00 p.m. on weekdays, with double staffing from noon to 2:00 p.m. each weekday, and single staffing the rest of the time. The desk was also single staffed on weekends for eight hours each day. Despite minimal staffing, desk staff experienced considerable downtime when there were no immediate questions to answer or technical issues to address. The consent form review service would be a way to optimize this downtime. Reference staff would be able to continue being productive while staffing the desk during less busy time periods.

Reference staff spent about twenty hours of their forty-hour workweek providing service at the reference desk. Their remaining work time was devoted to other regular duties, such as serving on library committees, attending meetings, training and supervising other staff members, planning and teaching library workshops, creating schedules, and contributing to various library projects. Although five reference staff members attended a health literacy training to prepare them to review research consent forms, only three of those five ended up providing the service. Of those three, one was a non-librarian staff member, and the other two were faculty-status librarians whose supervisory and teaching duties left little free time in their schedules.

With only three people available to review consent form submissions, the challenges of delegating the work and managing deadlines became evident early on. The online request form for the review service included a brief description of the service along with an assurance that "We will return the edited form to you within 3 business days." The form also contained a Review Requirements section that allowed requesters to select a "Need-By Date" and gave them the option to include "Additional Notes or Comments," if desired. Requesters could choose a date as early as the same day the request was created. This option led to a number of "rush" requests where researchers asked for same-day or next-day service. When the allotted timeframe is untenable because of previous obligations or staffing shortages at the desk, we have had to negotiate with the researchers for more time. We have since extended the turnaround time to five business days, but the "Need-By Date" remains on the form.

The consent form reviewers also noticed that the review submissions tended to increase or decrease in frequency according to the IRB meeting schedule. Within a week or two of a meeting, the IRB would issue deferral notice letters to principal investigators whose studies the IRB had not approved. When the deferral notice cited consent form language as an issue, some researchers would turn to the library's review service for help. As a result, the service would periodically receive multiple requests within one week, sometimes as many as three in one day.

To understand the time required to improve the readability of consent documents, it will help to compare the reading grade levels of those documents with the target reading level set by the university's Human Research Protections Office (HRPO). As we noted in our 2014 article about the service, HRPO began in 2009 to require that research consent forms be written at a seventh-grade reading level, a "stringent requirement" that was rarely met.[12] In fact, the average reading grade level of consent forms submitted to the library's review service since July 2011 has been 10.5. Library reviewers have managed to reduce the reading grade level of these documents by 2.2 grade levels on average, for an average post-review reading grade level of 8.4. Since the seventh-grade target is so rarely met, an eighth- to ninth-grade level is usually acceptable to the IRB.

While reducing the reading grade level by multiple grades requires significant editing and time commitment, document length is another important factor when considering editing time. We have reviewed consent documents with as few as three and as many as forty-five single-spaced pages, but the median submission length is 11.5 pages. Realizing that staff were spending considerable time on the service, in 2018, we began to track the total editing time for each review. As of December 2020, the average editing time per form was 7.3 hours.

Adding to the time challenge of providing the review service, the library merged its service desks in summer 2015. As part of this transition to a single-service-point model, the circulation and reference service staff was reduced; 2.5 full-time positions were eliminated, and the two departments merged to form the Information Services department. Staff working at the Information Services desk provide all walk-in, non-scheduled services for the library, including reference and research support, book and equipment lending, printing assistance, technical troubleshooting, cash transactions, and poster printing and sales. The desk is double-staffed from noon to 3:00 p.m. each day and single-staffed the rest of the time.

Although this desk model affords much less downtime for working on projects at the desk, the constraint of the researchers' "need-by dates" often necessitates that we use every available opportunity to complete the reviews on time. When the desk is busy, reviewers must juggle opposing demands: doing the intricate work of editing and reworking sentences while addressing a steady stream of service requests that may involve checking a computer connection on an upper floor of the library, to removing a paper jam from a public printer, or to demonstrating how to run a PICOT search in PubMed. When staff must toggle between editing work and customer service, each task disrupts and compromises the other. The distraction of service demands slows the editing work, while the level of focus required for editing conflicts with customer service standards that staff should appear available and approachable.

TEMPLATES AND SPONSOR-WRITTEN FORMS

Library reviewers have seen consent forms grow longer in response to an increasing number of informed consent regulations and requirements on the federal, state, and institutional level. These requirements can be complicated and have only become more so after the Common Rule, the federal policy that specifies the general requirements for informed consent, was updated in 2018. Linda Aldoory, director of the Hershel S. Horowitz Center for Health Literacy at the University of Maryland, College Park, has observed that these regulations introduce constraints that can "interfere with our ability to create health-literate informed consent."[13] Research institutions and study sponsors often rely on templates to guide consent form development, and to ensure compliance with federal, state, and institutional regulations governing the informed consent process. These templates impose constraints by mandating specific language that must be included in the form and may not be altered. The mandated language frequently includes instances of legalese or other high-grade-level bureaucratic writing.

Until recently, the UMB IRB would only review consent forms that conformed to either the UMB template, or a similar one for research conducted at the nearby Baltimore Veterans Affairs (VA) Medical Center; forms based on a third-party sponsor template would not be accepted. To comply with this mandate, UMB researchers collaborating on multi-center clinical trials sponsored by large pharmaceutical or biotech companies often attempt to satisfy both sponsor and university requirements by taking content from the sponsor-written ICF and inserting it into the university template using the copy-and-paste tools in Word.

Unfortunately, in some of the forms we have reviewed, we see evidence that entire blocks of text have been transferred from the sponsor form, with little care taken to incorporate the information into its new context. If the sponsor template uses different section headings, researchers may even paste sponsor content into the wrong section of the university's template. More commonly, they will add sponsor content that duplicates information already covered in the template. This strictly additive approach to combining two documents creates considerable repetition when a single element of consent, such as the right to withdraw, is addressed twice—first in language by the sponsor, and again within the university's template language. The resulting redundancy impairs readability in two ways. It creates a longer, wordier document, which is apt to be more off-putting to the reader. It also restates information in slightly different language, which is apt to confuse readers who must compare different iterations and wonder which to rely upon.

IRB reviewers have begun to flag redundancy and wordiness as factors that impair consent form readability. In their deferral notices to the principal investigators, the IRB has recommended the following for forms with readability issues:

> Please revise the ICF so it is easier for participants to understand. For assistance with simplifying the language to make the consent more understandable to the subject, please submit the consent document for this study to the UMB Health Sciences and Human Services Library for review, [web link].

When the readability issues cited include redundancy and length, library reviewers are expected to resolve these problems in addition to dealing with language issues on the sentence level. This involves synthesizing sponsor and UMB template content, eliminating redundant information, and ensuring that each template section contains only the required information. Both investigators and sponsors tend to err on the side of inclusion when it comes to informed consent, presumably believing that greater disclosure will help them meet informed consent requirements and avoid future liability. In practice, however, adding nonessential, redundant, or overly detailed information can overwhelm readers and detract from the information that is essential to informed consent.

In addition to the problems that sponsor templates introduce, there are problems inherent to the UMB and VA templates themselves. According to information posted on the HRPO website, the reading level target for consent forms is a seventh-grade level for UMB forms and an eighth-grade level for VA forms.[14] Like many other research institutions, however, UMB and the Baltimore VA provide ICF templates with readability scores considerably higher than their stated targets. The actual reading grade level of the UMB and the Baltimore VA templates are 12.1 and 12.3, respectively.

Dr. Michael Paasche-Orlow, a researcher and health literacy expert at Boston University School of Medicine, coined the term "hypocrisy index" to describe the difference between the target grade level and the actual reading grade level of the template.[15] In a 2003 study, Dr. Paashe-Orlow identified sixty-one U.S. medical schools that specified target ICF reading grade levels on their websites, and he then compared the target level for each school with the actual reading grade level of the school's ICF template. Only five of the sixty-one met their stated target, with the remaining schools exceeding their target by 2.8 grade levels on average.[16] By comparison, the UMB and Baltimore VA ICF templates score considerably higher than average on the hypocrisy index scale, at 5.1 for the nine-page UMB template and 4.3 for the twenty-one-page VA template.

While the templates also contain a mix of instructive and explanatory content, the majority of the content is composed of required or suggested statements. Instructions direct that one should, "for example, state if applicable:" or "state the following in this section." Readability challenges arise when the required or suggested statements are written in legalese or in a formal, bureaucratic style. A quick sampling of four single-sentence statements from the UMB ICF template yielded grade level scores of 13.4, 16.4, 16.7, and 21.1. Occasionally, when library reviewers have paraphrased these high-level statements to simplify them, we have encountered pushback from research specialists or HRPO staff who maintain that any statement from the template that appears within quotes is inviolable and cannot be altered in any way. However, whether the statements are required to remain verbatim appears to be a matter of interpretation. Some of the statements are inappropriate without modification, while others are misleading, or even grammatically incorrect.

Consent form submissions have also grown longer after January 21, 2019, when the revised 2018 Common Rule went into effect.[17] The 2018 rule contains additional elements of informed consent including the provision of a concise summary containing "key information" about the study, and, when relevant, statements regarding biospecimen collection and use, future research, clinically relevant results, and whole genome sequencing.[18] The purpose of the concise summary is to highlight key aspects of the study—purpose, procedures, risks, benefits—at the very beginning of the ICF; this is meant to give the participant or legally authorized representative a sense of whether they may want to take part in the study before they commit to reading the entire consent form.

With more complex studies, however, we have noticed that researchers may have trouble determining what information is key for the prospective participant's decision making. Again, the tendency is to err on the side of inclusion. Several of the submissions we have reviewed contain concise summaries that span three pages. While the 2018 rule provides no guidance on what qualifies as "concise," it is doubtful that adding another three pages of reading to a twenty-five-page form will improve the informed consent process from the participant's perspective. In our reviews, we have offered distilled versions of these summaries for the researchers' consideration, although we are arguably less well equipped to summarize the research than are the researchers themselves, who because they have designed the research study and written the protocol have a more thorough understanding of the research.

OUTSIDER STATUS AND LACK OF COMMUNICATION

If we were able to establish a dialogue with the researchers, it might help to resolve some of the questions that arise as we attempt to streamline a concise summary or clarify a set of study procedures. Unfortunately, the review service does not involve collaboration between the reviewer and the researcher or research team. The researchers simply submit the forms online, and we return the edited versions to them by email. It is a purely transactional exchange, performed at arm's length. There is no initial consultation meeting, no chance to discuss elements of the study that might be unclear, no opportunity to weigh different approaches to resolving problems. As a result, we must rely upon our own judgement when making changes to the ICF. If we have questions for the researchers, or are unsure if we have described something accurately, we can add margin comments to address these concerns.

In this way, the review service is essentially a third-party service, where the library reviewers have outsider status as nonscientist university staff who are involved in neither conducting nor monitoring the research. As outsiders, we are able to see the research from a layperson's perspective, which can be beneficial when trying to explain a research study to a lay audience. Yet this outsider perspective can also present challenges when we have to explain medical conditions and procedures that are, at times, entirely unfamiliar to us. Simplifying technical language requires a good deal of rephrasing. It is acceptable, and often unavoidable, to sacrifice some degree of precision to achieve a simpler phrasing; however, it is unacceptable to sacrifice accuracy.

As reviewers, we do our best to maintain accuracy by incorporating a lot of ad-hoc research in our review process. This may involve Googling various medical conditions or physiological processes; comparing how different consumer health websites describe a particular clinical procedure; or cross-referencing adverse effects or warnings for certain drugs or devices.

Another consequence of our outsider status is that researchers may develop misconceptions about the nature of the review service, how reviews are done, what falls outside the scope of the service, and what qualifications and training we have as reviewers. Based upon some of the quick turnaround times requested, some requesters may assume we have dedicated reviewers on staff whose sole job is to perform reviews. Alternatively, they may assume that reviewing a twenty-five-page ICF for a study of a new neurosurgical technique and lowering its reading level by four grade levels should be a relatively quick and easy task. They may even think we have access to some kind of special software that can facilitate the reviews. Since the service is impersonal and remote, and we are neither colleagues nor collaborators with those who request it, the requester's focus is likely on an IRB submission deadline; if they have thought about the service at all, it is only as a means of getting past an immediate barrier. Whatever the reason, we have had to find ways to manage expectations. For matters of time, this may involve telling the requester that we are short-staffed at the library on a particular week and need to balance the review work with service desk coverage and other obligations.

There are also misconceptions about the scope of the service and our qualifications as reviewers. In addition to the ICF document, some researchers will submit with their request an IRB deferral letter or a study protocol. They will then note in the request that they want us to address and resolve the "reasons for deferral" as itemized in the letter. When the deferral letter mentions discrepancies between the study procedures described in the ICF and those laid out in the official study protocol, the requester may ask us to refer to the protocol to resolve any issues. We consider it outside the scope of our service to review both the protocol and the ICF to ensure that the one is an accurate reflection of the other.

When the library introduced the consent form review service in 2011, the purpose and scope of the service was defined by a single statement: *We will review your consent form and make comments and suggested edits to help ensure that it is at an appropriate reading level for study participants.* That brief statement remains as the only description of what the service entails and why it is offered. As a definition, it is notably vague: What kind of "comments and suggested edits" will be made? And how will those comments and edits "ensure that [the ICF] is at an appropriate reading level for study participants"? The definition statement appears on a page of the HRPO website that contains links to the various informed consent and assent templates. Just below, the phone number for the library's Information Services desk is provided "for questions or concerns." Nowhere is there a link to a full description of the service because such a description does not exist, either as a public or internal document.

The fact that the service is not clearly delineated and is advertised on the HRPO site in a way that blurs the division between two separate entities—the HRPO and the library—gives the impression that HRPO is involved in and sanctions the service. This has helped to create a number of misconceptions among the researchers and research staff, some of whom seem to believe that the library reviewers have training as human research protections specialists and that one can call the library's Information Services desk for answers to questions about HRPO policies and requirements.

On two occasions, we asked HRPO staff to remove the "questions and concerns" contact information from their webpage, but our requests never made it to their webmaster. Because our relationship with the HRPO is more as a foreign entity than a collaborator, we have little control over how information about our service appears on the HRPO website. We continue to seek a contact at HRPO who will listen and respond to our concerns.

While we may not be able to change how other entities on campus characterize our service, we do have the ability to define our service in a way that clearly delineates its purpose, scope, and methods. Looking back over the history of the service, we may well ask ourselves why we have allowed the service to continue for so long under such a vague definition. When we first envisioned the service, our focus was on the ICF language at the level of sentences and words. We would reduce the reading level by shortening sentences and simplifying their structure, and by substituting shorter, more common words for longer, more technical ones. At the time, we were not considering other, higher-level, literacy issues we might encounter, such as whether the document presents information in a logical order, whether it avoids redundancy, and whether formatting is used in a way that makes the document more (rather than less) accessible. When not handled properly, each of these issues can compromise the overall readability of the document.

It is fair to say that our service is not clearly defined because our vision for the service was unclear from the outset. At the time, we knew of a number of studies the IRB had repeatedly deferred, citing problems with ICF language each time. If the research teams were not able to revise the forms and resolve these problems, maybe the library could do something to help—it might not be perfect, but it would be better than nothing. So the service was introduced as a stopgap measure, filling an immediate need where no other options existed.

The library reviewers receive minimal training—one forty-five-minute session on health literacy and ICF best practices led by a librarian specializing in health literacy—and no formal standards, guidelines, or checklists to follow during the review process. Furthermore, no outcome has ever been established as a goal of the review process, aside from "help[ing to] ensure . . . an appropriate reading grade level." Our casual, unstructured approach in providing the service has created a situation where there is no way for us to promote consistency in the service or product we provide; the reviews we produce will inevitably vary based on the subjective judgement and preferences of each individual reviewer. Without consistency, standards, or outcome targets, how can we assess the product we are delivering? And how can we gauge if the service has succeeded in helping to ensure an appropriate reading grade level? The only tools we have to measure the impact of our edits are Microsoft Word's readability algorithms. Aside from the data these formulas yield, the only way we can appraise the success of the service is through the subjective assessments of the reviewer (*I think I did a good job.*) and the requester (*This is great! Thanks for your help.*).

BENEFITS OF THE SERVICE

In spite of all of its challenges, the service has been beneficial in several ways. The HRPO values the service highly and often recommends or even mandates use of the service to researchers whose consent forms were not readable enough to be approved during IRB review. Although one or two librarians serve as nonscientist members of the IRB at any given time, they are not able to undertake major revisions of consent forms in their capacity as IRB reviewers. The library service fills this gap by having a separate group of nonscientists dedicated to reviewing and editing just the ICFs as opposed to the full study protocols. This team brings a layperson's perspective to the IRB by reviewing the ICFs. They are able to visualize what parts of the form would confuse the average person and highlight areas with suggestions for revision.

Another benefit is the satisfaction of the researchers who have used the service, which has been measured both quantitatively and qualitatively. The number of submissions to the service has steadily increased over the years since its inception, with a current average submission rate of four per month. Additionally, 100 percent of people who completed our evaluation survey after receiving a reviewed ICF reported that they would recommend the service to colleagues. Satisfied researchers have also praised the service through email and the survey, with many "repeat customers" submitting their ICFs for several subsequent studies. For example: "Thank you so very much for your time invested into improving this document. I am truly thankful for your help." "Very helpful!" and "This is amazing feedback. I am definitely going to be using this information in the future."

In addition to the benefits to the UMB research community, providing the review service is also rewarding for the library staff. Reviewing ICFs allows library staff to use their expertise in writing, editing, and communication in a unique way that adds variety to their work. Some of the staff have advanced degrees related to writing, while others simply have a background in the liberal arts and an interest in health literacy. The review service adds variety to their daily work.

Finally, a potential benefit that is more difficult to measure is the impact on the safety of UMB research study participants. Since ICFs are ultimately used to enroll people into research studies, improving the readability of these forms could help participants understand the risks and benefits involved. Many of the clinical trials conducted at UMB are extremely complex and involve a considerable number of safety risks, and the consent forms for these trials often include a lot of medical and scientific terminology. When these terms are adequately explained, in a way nonscientists can understand, people are empowered to make a well-informed choice about whether or not to take part in the research (see figure 10.2).

WHAT WE'VE LEARNED AND FUTURE DIRECTIONS

When we launched the review service ten years ago, we assumed that improving consent form readability would be relatively straightforward. Since the forms were so often written at high grade levels, using medical terminology and scientific research jargon, we would just "simplify" the instances of complexity by swapping out jargon and obscure terms and replacing them with lay language. We would shorten the sentences, use words with fewer syllables, add bulleted lists, and present information visually when possible, using tables or figures. As the service has progressed, we have realized there is a lot more to research consent form readability than basic health literacy guidelines can address.

Depending on the research study involved, informed consent forms can be extremely complex documents, with multiple stakeholder authors/editors, complicated study protocols, elaborate experimental designs, and an ever-growing, ever-changing list of regulatory requirements to consider. These aspects of the research are integral to the informed consent process and cannot be separated from the consent form language.

Despite the benefits of the service to date, the challenges involved are making it increasingly unsustainable in its current form. At this point, we are considering actions we can take to prevent having to discontinue the service altogether, such as training additional staff and limiting the scope of the reviews we provide. For example, we may be able to recruit graduate students interested in health literacy to serve as reviewers, which would lessen the need to try and recruit additional overworked library staff members. We are also contemplating offering training sessions and one-on-one consultations to researchers about plain language and readability as an alternative to reviewing full consent documents.

One idea that we have broached with the HRPO is to update the template language provided to investigators. Like many other institutions, UMB's ICF template language does not meet the readability standards recommended by the HRPO, which puts investigators in a somewhat impossible situation.[19] If they are not allowed to modify the template language—which in many cases appears to serve more as liability protection for the institution rather than information for participants—the researchers may be unable to meet the suggested readability level. If HSHSL librarians are able to work with the HRPO to edit the template language for greater readability, it could potentially improve many UMB consent forms and ultimately reduce some of the burden on the review service. One institution that revised their ICF template in this manner saw a great improvement in the readability level of investigator-initiated ICFs, which indicates that this approach can be successful.[20]

Another idea involves a more extreme change, as it would essentially transfer the service from the library to the HRPO. Given the many challenges library staff face when reviewing consent forms, it may be more appropriate for such review to be done by trained professionals. When Hadden et al. developed a more readable template at their institution, their team included research participants, a research ethicist, their Center for Health Literacy, and their Translational Research Institute.[21] Ethicists and health literacy experts might be ideal personnel to run a consent form review service. If our library's service reaches a point where it can no longer be sustained by library staff, an alternative would be for our HRPO to take the reins and involve dedicated, trained staff members with backgrounds in ethics or health literacy in running the service.

Ultimately, we do hope to continue offering this valuable service to the UMB community. To address the more immediate sustainability concerns, we can train additional staff to participate in reviewing materials and outline clearer expectations for what is and is not in the scope of the reviews. Creating a clear, detailed definition of the service would be especially useful because it would ensure researchers' expectations match those of the reviewers. Additionally, if we begin offering more training opportunities for researchers to learn about plain language writing and ICF readability, the quality of ICFs produced by UMB researchers may improve, reducing the need for complicated, time-consuming reviews. Lastly, improving our relationship with the HRPO will help us to address some of the challenges related to the university ICF templates and how the service is understood by the IRB. Recent conversations with HRPO staff indicate a willingness to consider changes, so we are optimistic that the ICF review service will be able to continue in some form into the future.

NOTES

1. Christine Grady, "Institutional Review Boards: Purpose and Challenges," *Chest* 148, no. 5 (November 2015): 1148–55, doi:10.1378/chest.15-0706.

2. Office for Human Research Protections, "Informed Consent FAQs," *HHS.Gov*, accessed July 24, 2020, https://www.hhs.gov/ohrp/regulations-and-policy/guidance/faq/informed-consent/index.html.

3. Ibid.

4. "General Requirements for Informed Consent," 45 CFR § 46.116 Electronic Code of Federal Regulations (eCFR) Title 45 § 46.116 (2018), https://www.ecfr.gov/cgi-bin/retrieveECFR?gp=&SID=83cd09e1c0f5c6937cd9d7513160fc3f&pitd=20180719&n=pt45.1.46&r=PART&ty=HTML#se45.1.46_1116.

5. Kevin Mertz et al., "The Reading Level of Surgical Consent Forms in Hand Surgery," *Journal of Hand Surgery Global Online* 1, no. 3 (July 1, 2019): 149–53, doi:10.1016/j.jhsg.2019.04.003.

6. Michael K. Paasche-Orlow, Holly A. Taylor, and Frederick L. Brancati, "Readability Standards for Informed-Consent Forms as Compared with Actual Readability," *New England Journal of Medicine* 348, no. 8 (February 20, 2003): 721–26, doi:10.1056/NEJMsa021212.

7. Mertz et al., "The Reading Level of Surgical Consent Forms in Hand Surgery"; Subha Perni et al., "Assessment of Use, Specificity, and Readability of Written Clinical Informed Consent Forms for Patients with Cancer Undergoing Radiotherapy,"

JAMA Oncology, May 2, 2019, e190260, doi:10.1001/jamaoncol.2019.0260; Laeeq Malik and James Cooper, "A Comparison of the Quality of Informed Consent for Phase I Oncology Trials over a 30-Year Period," *Cancer Chemotherapy and Pharmacology* 82, no. 5 (2018): 907–10, doi:10.1007/s00280-018-3673-x.

8. Malik and Cooper, "A Comparison of the Quality of Informed Consent for Phase I Oncology Trials over a 30-Year Period."

9. Kristie B. Hadden et al., "Improving Readability of Informed Consents for Research at an Academic Medical Institution," *Journal of Clinical and Translational Science* 1, no. 6 (April 23, 2018): 361–65, doi:10.1017/cts.2017.312.

10. "IRB Research Consent Form Review," *University of Maryland Health Sciences and Human Services Library*, accessed July 31, 2020, https://www2.hshsl .umaryland.edu/hshsl/assistance/research-consent.cfm.

11. "Consent and Assent Form Templates," *Human Research Protections*, accessed July 31, 2020, https://www.umaryland.edu/hrp/for-researchers/consent-form -templates/.

12. Paula G. Raimondo et al., "Health Literacy and Consent Forms: Librarians Support Research on Human Subjects," *Journal of the Medical Library Association* 102, no. 1 (January 2014): 5–8, doi:10.3163/1536-5050.102.1.003.

13. Institute of Medicine, *Informed Consent and Health Literacy: Workshop Summary* (Washington, DC: The National Academies Press, 2015), doi:10.17226/19019.

14. "Tools and Tips for Writing a Clear and Appropriate Informed Consent Document," 2006, https://www.umaryland.edu/media/umb/oaa/hrp/documents/study -tools-docs/clearconsent.pdf; VA Maryland Health Care System, "VA Informed Consent and HIPAA Guidelines," 2019, https://www.umaryland.edu/hrp/for-researchers/ consent-form-templates/.

15. Paasche-Orlow, Taylor, and Brancati, "Readability Standards for Informed-Consent Forms as Compared with Actual Readability."

16. Ibid.

17. Office for Human Research Protections, "Revised Common Rule," *HHS.Gov*, January 17, 2017, https://www.hhs.gov/ohrp/regulations-and-policy/regulations/fi nalized-revisions-common-rule/index.html.

18. "General Requirements for Informed Consent."

19. Paasche-Orlow, Taylor, and Brancati, "Readability Standards for Informed-Consent Forms as Compared with Actual Readability."

20. Hadden et al., "Improving Readability of Informed Consents for Research at an Academic Medical Institution."

21. Ibid.

BIBLIOGRAPHY

"Consent and Assent Form Templates." *Human Research Protections*. Accessed July 31, 2020. https://www.umaryland.edu/hrp/for-researchers/consent-form-tem plates/.

General requirements for informed consent, 45 CFR § 46.116 Electronic Code of Federal Regulations (eCFR) Title 45 § 46.116 (2018). https://www.ecfr.gov/cgi-bin/retrieveECFR?gp=&SID=83cd09e1c0f5c6937cd9d7513160fc3f&pitd=20180719&n=pt45.1.46&r=PART&ty=HTML#se45.1.46_1116.

Grady, Christine. "Institutional Review Boards: Purpose and Challenges." *Chest* 148, no. 5 (November 2015): 1148–55. doi:10.1378/chest.15-0706.

Hadden, Kristie B., Latrina Y. Prince, Tina D. Moore, Laura P. James, Jennifer R. Holland, and Christopher R. Trudeau. "Improving Readability of Informed Consents for Research at an Academic Medical Institution." *Journal of Clinical and Translational Science* 1, no. 6 (April 23, 2018): 361–65. doi:10.1017/cts.2017.312.

Institute of Medicine. *Informed Consent and Health Literacy: Workshop Summary.* Washington, DC: The National Academies Press, 2015. doi:10.17226/19019.

"IRB Research Consent Form Review." *University of Maryland Health Sciences and Human Services Library.* Accessed July 31, 2020. https://www2.hshsl.umaryland.edu/hshsl/assistance/research-consent.cfm.

Malik, Laeeq, and James Cooper. "A Comparison of the Quality of Informed Consent for Phase I Oncology Trials over a 30-Year Period." *Cancer Chemotherapy and Pharmacology* 82, no. 5 (2018): 907–10. doi:10.1007/s00280-018-3673-x.

Mertz, Kevin, Matthew B. Burn, Sara L. Eppler, and Robin N. Kamal. "The Reading Level of Surgical Consent Forms in Hand Surgery." *Journal of Hand Surgery Global Online* 1, no. 3 (July 1, 2019): 149–53. doi:10.1016/j.jhsg.2019.04.003.

Office for Human Research Protections. "Informed Consent FAQs." *HHS.Gov.* Accessed July 24, 2020. https://www.hhs.gov/ohrp/regulations-and-policy/guidance/faq/informed-consent/index.html.

Office for Human Research Protections. "Revised Common Rule." *HHS.Gov*, January 17, 2017. https://www.hhs.gov/ohrp/regulations-and-policy/regulations/finalized-revisions-common-rule/index.html.

Paasche-Orlow, Michael K., Holly A. Taylor, and Frederick L. Brancati. "Readability Standards for Informed-Consent Forms as Compared with Actual Readability." *New England Journal of Medicine* 348, no. 8 (February 20, 2003): 721–26. doi:10.1056/NEJMsa021212.

Perni, Subha et al., "Assessment of Use, Specificity, and Readability of Written Clinical Informed Consent Forms for Patients with Cancer Undergoing Radiotherapy." *JAMA Oncology* (August 8, 2019). e190260, doi:10.1001/jamaoncol.2019.0260.

Raimondo, Paula G., Ryan L. Harris, Michele Nance, and Everly D. Brown. "Health Literacy and Consent Forms: Librarians Support Research on Human Subjects." *Journal of the Medical Library Association* 102, no. 1 (January 2014): 5–8. doi:10.3163/1536-5050.102.1.003.

"Tools and Tips for Writing a Clear and Appropriate Informed Consent Document," 2006. https://www.umaryland.edu/media/umb/oaa/hrp/documents/study-tools-docs/clearconsent.pdf.

Chapter Eleven

Introduction to Reporting Guidelines Used in Animal Research

Effectiveness and Advocacy

Melissa C. Funaro and Kate Nyhan

A reporting guideline or standard is a simple, structured tool that provides a minimum list of information needed to ensure the replicability and transparency of research. Overall, following guidelines when reporting or preparing for animal experiments will increase the quality of the literature since researchers can replicate experiments and build upon previously published work.[1] Researchers can use guidelines while writing or preparing to write their manuscripts, and such guidelines have been developed in many fields of research, including animal research.[2] (See table 11.1 for a list of guidelines for animal research.) An uptake of reporting guidelines by researchers can improve the quality, reproducibility, and translatability of studies involving research on animals. Additionally, if experiments are conducted and reported according to best practices, there is the potential of both improving animal welfare and the translation of animal research to human health.[3]

The Animal Research: Reporting *In Vivo* Experiments (ARRIVE) guidelines were developed by the National Centre for the Replacement Refinement and Reduction of Animals in Research (NC3Rs) were published in *PLoS Biology* in 2010,[4] and updated in 2019.[5] The goal of ARRIVE is to help improve the transparency and reproducibility of biomedical research and consists of a checklist of "essential items" to include in the manuscript and a "recommended set" that adds context to the study.

The Gold Standard Publication Checklist (GSPC) was developed in 2010 and published in the journal *Alternatives to Laboratory Animals* (*ATLA*) by researchers at the Central Animal Laboratory and 3R Research Centre in the Netherlands.[6] Similar to the ARRIVE guideline, the purpose of the GSPC is to enable researchers to replicate and build upon research. However, a further

Table 11.1. List of Animal Guidelines and Their Uses

Guideline	Description
Animal Research: Reporting *In Vivo* Experiments (ARRIVE) 2019 (update of ARRIVE 2010)	Checklist designed to be used when submitting manuscripts describing animal research.
Gold Standard Publication Checklist (GSPC) 2010	Some overlap with ARRIVE, asks for more detail in husbandry techniques for planning, designing, and performing animal experiments
Guidance for the Description of Animal Research in Scientific Publications 2011	Guidance for effective reporting of animal research
Minimal information for publication of experimental pathology data (MINPEPE) 2016	Builds on ARRIVE plus minimum information needed to allow assessment of pathology data gathered from animal tissues
Strengthening the Reporting of Observational Studies in Epidemiology - Veterinary (STROBE-Vet) Statement 2016	Modification of Strengthening the Reporting of Observational Studies in Epidemiology (STROBE) Statement. Guidance for the reporting of observational studies related to animals. Guidance for the reporting of observational studies related to animals
Randomized Controlled trials for Livestock and Food Safety (REFLECT) statement 2010	Modification of the CONsolidated Standards of Reporting Trials (CONSORT) statement for reporting randomized controlled trials in human medicine
Planning Research and Experimental Procedures on Animals: Recommendations for Excellence (PREPARE) 2018	Guideline for planning animal research

goal of GSPC is to make future systematic reviews and meta-analyses of animal studies feasible.

The minimal information for publication of experimental pathology data (MINPEPE) guidelines[7] were published in 2016. These guidelines build upon the ARRIVE guidelines adding information requirements specific to experiments that gather data from animal tissues.

The Guidance for the Description of Animal Research in Scientific Publications[8] was created in 2011 by the Committee on Guidelines for Scientific Publications Involving Animal Studies. The report complements the ARRIVE and the GSPC guidelines, but in addition it includes specific recommendations for studies using aquatic species.

Strengthening the Reporting of Observational Studies in Epidemiology-Veterinary (STROBE-Vet) statement[9] was developed as an extension to the STROBE (Strengthening the Reporting of Observational Studies in Epidemiology) statement by an international steering committee of experts familiar with observational studies on animals. The goal of STROBE-Vet is to provide author guidance in describing their experiments and improve the reporting of observational studies in veterinary medicine related to health, production, welfare, and food safety.

The Randomized Controlled Trials for Livestock and Food Safety (RE-FLECT) statement[10] are guidelines designed for reporting randomized controlled trials in livestock and food safety. The goal of the guideline is to improve the quality of reporting and design of livestock trials and to increase reproducibility and transparency in the field.

In contrast to the aforementioned guidelines, the Planning Research and Experimental Procedures on Animals: Recommendations for Excellence (PREPARE) guidelines help researchers plan animal experiments[11] and serve as a complement to reporting guidelines such as ARRIVE and the GSPC. Which reporting guideline a researcher chooses to use will depend on their type of research and study design. More guidelines for specific types of experiments can be found through the Enhancing the QUAlity and Transparency Of health Research (EQUATOR) network (www.equator-network.org) or the Menagerie of Reporting guidelines Involving Animals (MERIDIAN) website (meridian.cvm.iastate.edu/).

WHY REPORTING GUIDELINES CAN INCREASE REPRODUCIBILITY AND TRANSPARENCY OF ANIMAL STUDIES AND ADVANCE THE IMPLEMENTATION OF THE 3RS

The use of reporting guidelines may help advance the implementation of the 3Rs in two ways. First, by increasing quality and reproducibility, fewer studies may need to be repeated and fewer animals will undergo distressful or painful procedures. And, secondly, using guidelines can help improve the search and retrieval of relevant articles when searching for alternatives because relevant information would have been included in the title or abstract.

A search for alternatives by investigators is required by the Animal Welfare Act.[12] The act requires that investigators provide an organization's Institutional Animal Care and Use Committee (IACUC) with documentation demonstrating that alternatives to procedures that may cause more than momentary pain or distress to animals have been considered and that activities do not unnecessarily duplicate previous experiments. Alternatives are often referred to as the 3Rs: the reduction in the number of animals used, refinement of techniques and procedures that reduce pain or distress, and replacement of animal with non-animal techniques. The 3Rs were introduced by Russell and Burch in 1959 in their publication *The Principles of Humane Experimental Techniques*[13] to guide the ethical use of animals in research. Russell and Birch propose that researchers should, where possible, reduce the number of animals used, refine their techniques to minimize pain, and replace animals with non-human models.

The United States Department of Agriculture (USDA), in enforcing the Animal Welfare Act, considers a search of the literature as "the most effective and efficient method for demonstrating compliance with the requirement."[14] However, a database search is no stronger than the literature in the field it searches,[15] and identifying relevant publications can be difficult if the literature doesn't contain relevant information. Since most databases mainly search the title and abstracts, if the relevant species or procedures aren't included in the title or abstract, these articles will be missed from a search no matter how comprehensive the search strategy.

Identifying and retrieving relevant articles proves even more difficult if important information is buried in the full text and not reported effectively in the abstract. The incomplete reporting of relevant details (such as species, procedures), within those publications can make a search strategy unhelpful.[16] Because only a small number of databases permit full-text searching of scholarly articles, a search strategy can be comprehensive in terms of vocabulary for species or procedures and still fail to retrieve relevant articles, especially if those relevant articles use the key words in the body but not the title or abstract. As researchers increasingly use reporting guidelines, it may become easier to identify papers about alternatives to painful or distressful procedures.

For a detailed description of alternatives searching, using the 3Rs approach, see chapter 6 in this book, "Librarians and the IACUC: Practical Approaches for Performing Alternatives Searches and Providing Support." Alternatively, in a hypothetical world where reporting guidelines are heavily used for animal research, another path beckons. Instead of retrieving and screening only those articles that refer to the procedure of interest explicitly in the title or abstract, a searcher could search for papers that (1) declare themselves compliant with reporting guidelines and (2) address the species and/or disease model of interest. The first requirement would screen out articles that do not include enough detail to be useful for identifying alternatives. This could reduce the screening workload of articles needed by researchers and the search for alternatives could retrieve more relevant results and not be seen as a burden.[17] Articles with detailed reporting, on the other hand, would be retrieved and screened even if they do not mention the procedure of interest in the title or abstract, as long as they address the species and/or disease model of interest.

Teaching Library Users about Reporting Guidelines for Animal Research

The 3Rs and IACUC searching are a topic of instruction as well as a mediated search service; reporting guidelines can similarly be a valuable topic in library instruction programs. In our experience, it is important to frame your discussion of reporting guidelines with a realistic understanding of animal researchers' incentives. When librarians show an understanding of researchers' priorities, goals, and pressures, it is possible to build trust and rapport. Trust and rapport are especially important in the context of animal reporting guidelines because the adoption of reporting guidelines for animal research, while not new, is not yet widespread.

Early career researchers who work with animals may see few examples of complete reporting in the literature they read.[18] When they read highly cited, well-regarded papers—indeed, seminal papers—that do not fully comply with reporting guidelines, why should *they* (the less well-known researchers) invest time and effort in reporting guideline compliance? Any library-IACUC instruction or consultation that ignores this context will be perceived by researchers as naïve and impractical. If librarians are perceived in this way, trust and rapport cannot be developed.

Since animal researchers will routinely read and cite poorly reported work, reporting guidelines could be perceived as a tick-box exercise or a bureaucratic obstacle to publishing. Librarians can counter that perception by introducing reporting guidelines as useful tools to make writing easier and to make research more rigorous. IACUC consultations and instruction thus present an opportunity for professional development and engagement with departments and early career researchers.

Arguments in favor of reporting guideline compliance must, in our experience, engage researchers by emphasizing practical benefits. At several points in a researcher's career, reporting guidelines offer real-world benefits to the authors who use them. Discussing these benefits explicitly in instruction and consultations has produced positive results in our experience. The following paragraphs aim to familiarize librarians with the conversations they can have with researchers in different career contexts about reporting guideline compliance.

- In the short term—at the scale of an author writing a single paper—reporting guidelines offer researchers a set of best practices, clearly articulated by domain experts. Such clear guidance is especially valuable for early career researchers but can be useful for authors at any career stage. Instead of attempting to synthesize or intuit best practices from the literature as a whole, or imitating a single published paper (which or may not be well reported), authors can refer to reporting guidelines and have confidence that they are meeting community-based standards.

• In the medium term—at the scale of a principal investigator leading a grant over several years—reporting guidelines can support smooth onboarding of new staff and trainees. If researchers are planning to fully report their work in the eventual publication, they will need to document their protocols in detail. The resulting detailed protocols are useful for onboarding and other transitions, even if they were originally drafted for the sake of future reporting guideline compliance.

• In the long term—at the scale of a career—reporting guideline compliance might have further benefits. If a lab's work is clearly reported, other labs can more easily build on it, potentially leading to new research that cites the publications by the original lab. Citations and successful replications can improve the reputation of the original lab, with multiple positive downstream effects, such as success recruiting top candidates, success placing trainees in prestigious fellowships, and success generating institutional funding. Conversely, external researchers might well be inspired by findings they read in a poorly reported paper, but the external researchers will likely have trouble replicating or extending it. If citations accrue at all, they might be slower and fewer. Indeed, external researchers might simply abandon their attempted replication or expansion when they are stymied by incomplete reporting in the original paper.

In instruction sessions, we often find that a participant can share a personal experience about reporting and republication; such real-life examples are always worth the time to discuss, because they debunk the notion that library-hosted reporting guideline instruction is naïve or impractical. These arguments in favor of complying with reporting guidelines all appeal to the self-interest of researchers, and librarians can use these as strategies for connecting effectively with graduate students, postdocs, and faculty seeking IACUC consultations and instruction at our institution.

There are also external reasons for researchers to engage with reporting guidelines. A growing number of journals state in their author instructions that they recommend or require reporting guideline compliance for animal research. Although reporting guidelines should not be used as quality evaluation tools unless the guideline was explicitly intended for that purpose,[19] peer reviewers can make use of reporting guidelines to increase the consistency of ratings and recommendations.[20] The ARRIVE 2.0 guidelines are explicated intended for the use of editors and reviewers as well as researchers.[21] NIH policies now emphasize the importance of rigor and reproducibility in training grants.[22] Animal research uses three scarce resources: animals, research funding, and researchers' time. Full reporting of animal research is good stewardship of these limited resources, and as such it is an ethical obligation.

Librarians are wise to make the case for reporting guidelines in animal research using all these arguments. It has been difficult for us to predict which rationale for the use of reporting guidelines will be effective with any individual researcher working with animals. Discussing multiple reasons to use reporting guidelines is thus part of our practice in consultations and instruction.

Many resources exist to support instruction around reporting guidelines; these resources are generally not limited to animal use reporting guidelines in particular but rather encompass reporting guidelines for different domains and research designs. The premier resource for reporting guidelines—published ones and ones under development—is the Equator Network (https://www.equator-network.org/). The Equator Network's Action Plan for institutions articulates practical suggestions about how to support researchers with reporting guidelines (https://www.equator-network.org/wp-content/uploads/2016/03/PAHO-Universities-Action-Sheet-31March2016_FINAL.pdf) and includes links to additional resources, including two Librarian Action Plans (https://www.equator-network.org/wp-content/uploads/2013/06/Librarian-Action-Plan-Simple-Ideas.pdf and https://www.equator-network.org/wp-content/uploads/2013/06/Librarian-Targeted-Actions.pdf). Librarians working with animal researchers can use these action plans to increase awareness and use of reporting guidelines. Among the action plan suggestions that we have implemented at our institution are these:

- mention reporting guidelines to researchers/faculty during consultations;
- promote reporting guidelines and how they can be easily used for planning and writing research papers;
- when providing the results of literature searches, advise researchers on the appropriate reporting guideline to use to write up their study; and
- work with faculty to ensure that reporting guidelines are included in course content.

The action plan suggestions will be most easily implemented at institutions where librarians are seen as expert partners with insight into scholarly communication. However, any librarian working with animal researchers can start to increase awareness and use of reporting guidelines, even if it is on a small scale. In our experience, early career researchers often have tangible, practical needs that reporting guidelines fill; some early career researchers become cheerleaders for reporting guidelines (and indeed the librarians who introduced them to these tools) and spread their new knowledge widely in labs and departments. Librarians can work with such early adopters to generate grassroots interest in reporting guidelines.

Library instruction about reporting guidelines can be situated in the larger context of reproducibility librarianship, a specialty of growing importance. Sayre and Riegelman discuss reporting guidelines in general and the ARRIVE guidelines in particular in their works on academic libraries' responses to the reproducibility crisis.[23] Some health sciences libraries have developed robust programs on research reproducibility; prominent among them is the University of Utah's Spencer S. Eccles Health Sciences Library, where their reproducibility agenda encompasses a new position with reproducibility and open science as a major component, a successful conference, trainings, short courses, and a Research Reproducibility Coalition that facilitates information exchange across campus.[24] Reproducibility librarianship encompasses computational reproducibility and aspects of data management; reporting guidelines are just one facet of this larger domain.[25] That said, nontransparent reporting of animal research was discussed in the keynote address at the online conference Librarians Building Momentum for Reproducibility, illustrating that reporting guideline awareness has a place in reproducibility librarianship.[26]

Returning to the topic of instruction on reporting guidelines for animal researchers, there are effective instruction techniques librarians can use in participatory trainings. We offer two approaches: starting with published articles or starting with the ARRIVE Guidelines 2.0.

To facilitate a session based on published articles, librarians can invite workshop participants to nominate articles about animal models that they have used in their own work. The criterion for nominations is not good reporting or bad reporting; instead, participants should focus on relevance to their research agenda. When the planned workshop is small, it might be wise to invite multiple nominations from each person.

Alternatively, the leader of a drop-in workshop could choose example articles; the article-level ARRIVE compliance dataset[27] could be a good source of example articles, or the leader could choose articles related to the research interests or animal models of expected workshop participants. Whether choosing articles from participants' suggestions or from the literature, it is important to ensure that the examples used in the workshop are not all completely compliant with reporting guidelines. At least some of the examples used in the workshop should be articles that are poorly reported (that is, not compliant with the ARRIVE Essential 10 checklist).

The selected example articles can be distributed to the workshop participants in advance, or the leaders can build reading time into the workshop, or both. In our experience, discussions are more active when we build reading time into the workshop, rather than relying on participants to read in advance. On the other hand, workshop participants who speak English as a foreign language may prefer the opportunity to prepare in advance.

In the workshop, participants can work through the articles with a reporting guideline checklist, such as the ARRIVE Essential 10, the ARRIVE Recommended Set, or another guideline from table 11.1. Key points to discuss will include the following:

- the degree of compliance,
- the reasons why some authors might not have included all the required or recommended elements, and
- whether the absence of some elements changes the participants' assessment of the article's usefulness and quality.

This instruction format—close reading of real articles in relation to real reporting guidelines—can be used for any type of reporting guideline, and it fits well into journal club settings as well as traditional training workshop settings. If time allows, workshop leaders can show the author instructions from the journals that published the example articles. The group can discuss whether a journal's author instructions and recommendations may have influenced the authors' use or non-use of reporting guidelines in the example papers. Another potential extension is to check for open referee reports, which are sometimes available on the journal website, and investigate whether any peer reviewers commented on the paper's compliance with any reporting guideline recommended in the journal's instructions to authors. Commentary from IACUC committee members or staff are another potential extension of this activity.

An alternative approach to workshop planning is to focus instead on the reporting guidelines themselves. The structure of the ARRIVE Guidelines 2.0, which are divided into "essential" and "recommended" sets, invites a card-sorting activity.[28] The card-sorting activity requires no advance preparation or advance engagement on the part of workshop participants.

The leader provides the participants—ideally, graduate students, postdocs, and staffers doing animal research—with the twenty-one ARRIVE items (*without* numbers or groupings) and prompts them to sort the items into groups: elements that you "need to have" in an article reporting animal research and elements that are "nice to have" in such an article. After participants have categorized the twenty-one items (working individually or in small groups), the leader introduces ARRIVE 2.0 and facilitates a discussion comparing the participants' decisions with the conclusions of the ARRIVE 2.0 experts. Any variance can be explored with reference to the ARRIVE 2.0 Explanation and Elaboration (E&E) document.[29] That document addresses each element of the essential and recommended set, providing models of compliance for each guideline item and justifying the importance of each item.

Leaders should be prepared to use the E&E to make the case that an "essential" item is "essential" even though participants did not consider it so (while also acknowledging that some authors decide against fully complying with reporting guidelines). Workshop leaders should also be prepared for participants to categorize as "essential" one of the officially "recommended" items; in that case it may be useful to remind participants that reporting guidelines are a floor, rather than a ceiling.

Fruitful discussion can arise when participants use their own experience in the lab to explain their card-sort decisions, and when participants disagree with the ARRIVE 2.0 guidelines in either direction. A prompt for further discussion is to invite participants to reflect on a recent animal model paper—from their favorite journal, using their favorite animal model, or from their own lab—and consider whether any ARRIVE items could have been better reported.

Whatever the format of an instruction session on reporting guidelines for animal research, there are two common misunderstandings that should be addressed explicitly. First, researchers should refer to reporting guidelines in the early stages of a project, not just during the process of preparing the manuscript. Reporting guidelines should be part of their planning process. A researcher who attempts to comply with a reporting guideline "after the fact" may find that they have not recorded all the necessary information and will waste precious time and effort trying to retrace their steps in order to document their search process. Second, it should be emphasized that reporting guidelines are neither methods guidelines nor critical appraisal guidelines,[30] even though reporting guidelines are sometimes incorrectly used as quality measures in meta-research.

Using the techniques and resources discussed in this chapter, librarians and other stakeholders will be able to share information about reporting guidelines with animal researchers at any career stage. We believe that increasing awareness of and compliance with reporting guidelines in animal research contribute to a reduction in research waste and improvements in reporting, reproducibility, and rigor.

NOTES

1. Douglas G. Altman and Iveta Simera, "Using Reporting Guidelines Effectively to Ensure Good Reporting of Health Research" (32–40), in David Moher, Douglas G. Altman, Kenneth F. Schulz, Iveta Simera and Elizabeth Wager (Eds.), *Guidelines for Reporting Health Research: A User's Manual* (Hoboken, NJ: Wiley, 2014); Stuart G. Nicholls, Sinéad M. Langan, Eric I, Benchimol, and David Moher, "Reporting Transparency: Making the Ethical Mandate Explicit," *BMC Medicine* 14, no. 1 (2016): 1–3; SeungHye Han, Tolani F. Olonisakin, John P. Pribis, Jill Zupetic, Joo Heung Yoon, Kyle M. Holleran, Kwonho Jeong, Nader Shaikh, Doris M. Rubio, and Janet S. Lee, "A Checklist Is Associated with Increased Quality of Reporting Preclinical Biomedical Research: A Systematic Review," *PLoS One* 12, no. 9 (2017): e0183591, doi: 10.1371/journal.pone.0183591.

2. Nathalie Percie du Sert, Viki Hurst, Amrita Ahluwalia, Sabina Alam, Marc T. Avey, Monya Baker, William J. Browne, et al., "The ARRIVE Guidelines 2019: Updated Guidelines for Reporting Animal Research," *bioRxiv* (2019): 703181, doi: 10.1101/703181; Carlijn R. Hooijmans, Marlies Leenaars, and Merel Ritskes-Hoitinga, "A Gold Standard Publication Checklist to Improve the Quality of Animal Studies, to Fully Integrate the Three Rs, and to Make Systematic Reviews More Feasible." *Alternatives to Laboratory Animals* 38, no. 2 (2010): 167–82; Adrian J. Smith, R. Eddie Clutton, Elliot Lilley, Kristine E. Aa Hansen, and Trond Brattelid, "PREPARE: Guidelines for Planning Animal Research and Testing," *Laboratory Animals* 52, no. 2 (2018): 135–41; Olga Giraldo, Alexander Garcia, and Oscar Corcho, "A Guideline for Reporting Experimental Protocols in Life Sciences," *PeerJ* 6 (2018): e4795; Cheryl L. Scudamore, Elizabeth J. Soilleux, Natasha A. Karp, Ken Smith, Richard Poulsom, C. Simon Herrington, Michael J. Day, et al., "Recommendations for Minimum Information for Publication of Experimental Pathology Data: MINPEPA Guidelines," *Journal of Pathology* 238, no. 2 (2016): 359–67; National Research Council (NRC), "Guidance for the Description of Animal Research in Scientific Publications" (2011).

3. Meredith Bara and Ari R. Joffe, "The Methodological Quality of Animal Research in Critical Care: The Public Face of Science," *Annals of Intensive Care* 4, no. 1 (2014): 1–9; Guilherme S. Ferreira, Désirée H. Veening-Griffioen, Wouter P. C. Boon, Ellen H. M. Moors, and Peter J. K. van Meer, "Levelling the Translational Gap for Animal to Human Efficacy Data," *Animals* 10, no. 7 (2020): 1199; Zahra Bahadoran, Parvin Mirmiran, Khosrow Kashfi, and Asghar Ghasemi, "Importance of Systematic Reviews and Meta-analyses of Animal Studies: Challenges for Animal-to-Human Translation," *Journal of the American Association for Laboratory Animal Science* 59, no. 5 (2020): 469–77; Carol Kilkenny, Nick Parsons, Ed Kadyszewski, Michael FW Festing, Innes C. Cuthill, Derek Fry, Jane Hutton, and Douglas G. Altman, "Survey of the Quality of Experimental Design, Statistical Analysis and Reporting of Research Using Animals," *PloS One* 4, no. 11 (2009): e7824; Larry Carbone, and Jamie Austin. "Pain and Laboratory Animals: Publication Practices for Better Data Reproducibility and Better Animal Welfare," *PloS One* 11, no. 5 (2016): e0155001.

4. Carol Kilkenny, William J. Browne, Innes C. Cuthill, Michael Emerson, and Douglas G. Altman, "Improving Bioscience Research Reporting: The ARRIVE Guidelines for Reporting Animal Research," *PLoS Biology* 8, no. 6 (2010): e1000412.

5. Percie du Sert et al., "The ARRIVE Guidelines 2019."

6. Hooijmans et al., "A Gold Standard Publication Checklist."

7. Scudamore et al., "Recommendations for Minimum Information."

8. National Research Council (NRC), "Guidance for the Description of Animal Research in Scientific Publications."

9. Jan M. Sargeant, Annette M. O'Connor, Ian R. Dohoo, Hollis N. Erb, Myriam Cevallos, Matthias Egger, Annette Kjær Ersbøll, et al. "Methods and Processes of Developing the Strengthening the Reporting of Observational Studies in Epidemiology–Veterinary (STROBE–Vet) Statement," *Journal of Veterinary Internal Medicine* 30, no. 6 (2016): 1887–95.

10. Annette M. O'Connor, Jan M. Sargeant, Ian A. Gardner, James S. Dickson, Mary E. Torrence, Cate E. Dewey, Ian R. Dohoo, et al., "The REFLECT Statement: Methods and Processes of Creating Reporting Guidelines for Randomized Controlled Trials for Livestock and Food Safety," *Journal of Food Protection* 73, no. 1 (2010): 132–39.

11. Smith et al., "PREPARE."

12. United States Department of Agriculture, *Animal Welfare Act of 1966* (PL 89-544) (approved August 24, 1966).

13. William Moy Stratton Russell, *The Principles of Humane Experimental Technique* (Springfield, IL: C.C. Thomas., 1959).

14. United States Department of Agriculture Animal Care, "Animal Care Resource Guide, Policy #12: Written Narrative for Alternatives to Painful Procedures," 2011.

15. Carbone and Austin, "Pain and Laboratory Animals."

16. Marc T. Avey, Nicole Fenwick, and Gilly Griffin, "The Use of Systematic Reviews and Reporting Guidelines to Advance the Implementation of the 3Rs," *Journal of the American Association for Laboratory Animal Science* 54, no. 2 (2015): 153–62.

17. National Institutes of Health, Animal and Plant Health Inspection Service, and Food and Drug Administration, "Reducing Administrative Burden for Researchers: Animal Care and Use in Research," 2019.

18. Vivian Leung, Frédérik Rousseau-Blass, Guy Beauchamp, and Daniel S. J. Pang, "ARRIVE has Not ARRIVEd: Support for the ARRIVE (Animal Research: Reporting of *In Vivo* Experiments) Guidelines Does Not Improve the Reporting Quality of Papers in Animal Welfare, Analgesia or Anesthesia," *PLoS One* 13, no. 5 (2018): e0197882.

19. Patricia Logullo, Angela MacCarthy, Shona Kirtley, and Gary S. Collins, "Reporting Guideline Checklists Are Not Quality Evaluation Forms: They Are Guidance for Writing," *Health Science Reports* 3, no. 2 (2020): e165.

20. Eunhye Song, Lin Ang, Ji-Yeun Park, Eun-Young Jun, Kyeong Han Kim, Jihee Jun, Sunju Park, and Myeong Soo Lee, "A Scoping Review on Biomedical Journal Peer Review Guides for Reviewers." *PLoS One* 16, no. 5 (2021): e0251440.

21. Nathalie Percie du Sert, Viki Hurst, Amrita Ahluwalia, Sabina Alam, Marc T. Avey, Monya Baker, William J. Browne, et al., "The ARRIVE Guidelines 2.0:

Updated Guidelines for Reporting Animal Research," *PLoS Biology* 18, no. 7(2020): e3000410.

22. National Institutes of Health (NIH), "Implementing Rigor and Transparency in NIH and AHRQ Career Development Award Applications," released 13 October 2015, https://grants.nih.gov/grants/guide/notice-files/NOT-OD-16-012.html.

23. Franklin Sayre and Amy Riegelman, "The Reproducibility Crisis and Academic Libraries," *College & Research Libraries* 79, no. 1 (2018): 2; Franklin Sayre and Amy Riegelman, "Replicable Services for Reproducible Research: A Model for Academic Libraries," *College & Research Libraries* 80, no. 2 (2019): 260.

24. Melissa L. Rethlefsen, Mellanye J. Lackey, and Shirley Zhao, "Building Capacity to Encourage Research Reproducibility and #MakeResearchTrue," *Journal of the Medical Library Association* 106, no. 1 (2018): 113–19.

25. Vicky Steeves, "Reproducibility Librarianship," *Collaborative Librarianship* 9, no. 2(2017): 80–89.

26. Melissa L. Rethlefsen, *Librarians and Reproducibility: It's Time* (2020). OSF Home. https://osf.io/u5b2f/.

27. Daniel Pang, "ARRIVE Guidelines Study Data" (*Harvard Dataverse, 2017*).

28. Percie du Sert et al., "The ARRIVE Guidelines 2.0."

29. Nathalie Percie du Sert, Amrita Ahluwalia, Sabina Alam, Marc T. Avey, Monya Baker, William J. Browne, Alejandra Clark, et al., "Reporting Animal Research: Explanation and Elaboration for the ARRIVE Guidelines 2.0." *PLOS Biology* 18, no. 7 (2020): e3000411.

30. Logullo et al., "Reporting Guideline Checklists Are Not Quality Evaluation Forms."

BIBLIOGRAPHY

Altman, Douglas G., and Iveta Simera. "Using Reporting Guidelines Effectively to Ensure Good Reporting of Health Research." In David Moher, Douglas G. Altman, Kenneth F. Schulz, Iveta Simera and Elizabeth Wager (Eds.), *Guidelines for Reporting Health Research: A User's Manual*, 32–40. Hoboken, NJ: Wiley. 2014.

Avey, Marc T., Nicole Fenwick, and Gilly Griffin. "The Use of Systematic Reviews and Reporting Guidelines to Advance the Implementation of the 3Rs." *Journal of the American Association for Laboratory Animal Science* 54, no. 2 (2015): 153–62.

Bahadoran, Zahra, Parvin Mirmiran, Khosrow Kashfi, and Asghar Ghasemi. "Importance of Systematic Reviews and Meta-analyses of Animal Studies: Challenges for Animal-to-Human Translation." *Journal of the American Association for Laboratory Animal Science* 59, no. 5 (2020): 469–77. doi: 10.30802/aalas-jaalas-19-000139.

Bara, Meredith, and Ari R. Joffe. " The Methodological Quality of Animal Research in Critical Care: The Public Face of Science." *Annals of Intensive Care* 4, no. 1 (2014): 1–9. doi: 10.1186/s13613-014-0026-8.

Carbone, Larry, and Jamie Austin. "Pain and Laboratory Animals: Publication Practices for Better Data Reproducibility and Better Animal Welfare." *PLoS One* 11, no. 5 (2016): e0155001. doi: 10.1371/journal.pone.0155001.

Ferreira, Guilherme S., Désirée H. Veening-Griffioen, Wouter P. C. Boon, Ellen H. M. Moors, and Peter J. K. van Meer. "Levelling the Translational Gap for Animal to Human Efficacy Data." *Animals* 10, no. 7 (2020): Article 1199. doi: 10.3390/ani10071199.

Giraldo, Olga, Alexander Garcia, and Oscar Corcho. "A Guideline for Reporting Experimental Protocols in Life Sciences." *PeerJ* 6 (2018): e4795. doi: 10.7717/peerj.4795.

Han, SeungHye, Tolani F. Olonisakin, John P. Pribis, Jill Zupetic, Joo Heung Yoon, Kyle M. Holleran, Kwonho Jeong, Nader Shaikh, Doris M. Rubio, and Janet S. Lee. "A Checklist Is Associated with Increased Quality of Reporting Preclinical Biomedical Research: A Systematic Review." *PLoS One* 12, no. 9 (2017): e0183591. doi: 10.1371/journal.pone.0183591.

Hooijmans, Carlijn R., Marlies Leenaars, and Merel Ritskes-Hoitinga. "A Gold Standard Publication Checklist to Improve the Quality of Animal Studies, to Fully Integrate the Three Rs, and to Make Systematic Reviews More Feasible." *Alternatives to Laboratory Animals* 38, no. 2 (2010): 167–82.

Kilkenny, Carol, William J. Browne, Innes C. Cuthill, Michael Emerson, and Douglas G. Altman. "Improving Bioscience Research Reporting: The ARRIVE Guidelines for Reporting Animal Research." *PLoS Biology* 8, no. 6 (2010): e1000412. doi: 10.1371/journal.pbio.1000412.

Kilkenny, Carol, Nick Parsons, Ed Kadyszewski, Michael F. W. Festing, Innes C. Cuthill, Derek Fry, Jane Hutton, and Douglas G. Altman. " Survey of the Quality of Experimental Design, Statistical Analysis and Reporting of Research Using Animals." *PLoS One* 4, no. 4 (2009):e7824. doi: 10.1371/journal.pone.0007824.

Leung, Vivian, Frédérik Rousseau-Blass, Guy Beauchamp, and Daniel S. J. Pang. "ARRIVE has Not ARRIVEd: Support for the ARRIVE (Animal Research: Reporting of *In Vivo* Experiments) Guidelines Does Not Improve the Reporting Quality of Papers in Animal Welfare, Analgesia or Anesthesia." *PLoS One* 13, no. 5 (2018): e0197882. doi: 10.1371/journal.pone.0197882.

Logullo, Patricia, Angela MacCarthy, Shona Kirtley, and Gary S. Collins. "Reporting Guideline Checklists Are Not Quality Evaluation Forms: They Are Guidance for Writing." *Health Science Reports* 3, no. 2 (2020): e165. doi: https://doi.org/10.1002/hsr2.165.

National Institutes of Health. "Implementing Rigor and Transparency in NIH and AHRQ Career Development Award Applications." Released October 13, 2015. https://grants.nih.gov/grants/guide/notice-files/NOT-OD-16-012.html.

National Institutes of Health, Animal and Plant Health Inspection Service, and Food and Drug Administration. "Reducing Administrative Burden for Researchers: Animal Care and Use in Research." 2019.

National Research Council (NRC). "Guidance for the Description of Animal Research in Scientific Publications." 2011.

Nicholls, Stuart G., Sinéad M. Langan, Eric I. Benchimol, and David Moher. "Reporting Transparency: Making the Ethical Mandate Explicit." *BMC Medicine* 14, no. 1 (2016): 1–3.

O'Connor, Annette M., Jan M. Sargeant, Ian A. Gardner, James S. Dickson, Mary E. Torrence, Cate E. Dewey, Ian R. Dohoo, et al. 2010. "The REFLECT Statement: Methods and Processes of Creating Reporting Guidelines for Randomized Controlled Trials for Livestock and Food Safety." *Journal of Food Protection* 73, no. 1 (2010): 132–39. doi: 10.1016/j.prevetmed.2009.10.008.

Pang, Daniel. "ARRIVE Guidelines Study Data." *Harvard Dataverse.* 2017.

Percie du Sert, Nathalie, Amrita Ahluwalia, Sabina Alam, Marc T. Avey, Monya Baker, William J. Browne, Alejandra Clark, et al. "Reporting Animal Research: Explanation and Elaboration for the ARRIVE Guidelines 2.0." *PLOS Biology* 18, no. 7 (2020): e3000411. doi: 10.1371/journal.pbio.3000411.

Percie du Sert, Nathalie, Viki Hurst, Amrita Ahluwalia, Sabina Alam, Marc T. Avey, Monya Baker, William J. Browne, et al. "The ARRIVE Guidelines 2.0: Updated Guidelines for Reporting Animal Research." *PLoS Biology* 18, no. 7 (2020):e3000410. doi: 10.1371/journal.pbio.3000410.

Percie du Sert, Nathalie, Viki Hurst, Amrita Ahluwalia, Sabina Alam, Marc T. Avey, Monya Baker, William J. Browne, et al. "The ARRIVE Guidelines 2019: Updated Guidelines for Reporting Animal Research." *bioRxiv* (2019):703181. doi: 10.1101/703181.

Rethlefsen, Melissa L., Mellanye J. Lackey, and Shirley Zhao. "Building Capacity to Encourage Research Reproducibility and #MakeResearchTrue." *Journal of the Medical Library Association* 106, no. 1 (2018): 113. doi: 10.5195/jmla.2018.273.

Rethlefsen, Melissa L. 2020. *Librarians and Reproducibility: It's Time.* OSF Home. https://osf.io/u5b2f/.

Russell, William Moy Stratton.*The Principles of Humane Experimental Technique.* Springfield, IL: C.C. Thomas. 1959.

Sargeant Jan M., Annette M. O'Connor, Ian R. Dohoo, Hollis N. Erb, Myriam Cevallos, Matthias Egger, Annette Kjær Ersbøll, et al. "Methods and Processes of Developing the Strengthening the Reporting of Observational Studies in Epidemiology-Veterinary (STROBE-Vet) Statement." *Journal of Veterinary Internal Medicine* 30, no. 6 (2016): 1887–95. doi: 10.4315/0362-028x.Jfp-16-016.

Sayre, Franklin, and Amy Riegelman. "The Reproducibility Crisis and Academic Libraries." *College & Research Libraries* 79, no. 1 (2018): 2.

Sayre, Franklin, and Amy Riegelman. 2019. "Replicable Services for Reproducible Research: A Model for Academic Libraries." *College & Research Libraries* 80, no. 2 (2019): 260.

Scudamore, Cheryl L., Elizabeth J. Soilleux, Natasha A. Karp, Ken Smith, Richard Poulsom, C. Simon Herrington, Michael J. Day, et al. "Recommendations for Minimum Information for Publication of Experimental Pathology Data: MINPEPA Guidelines." *Journal of Pathology* 238, no. 2 (2016): 359–67. doi: 10.1002/path.4642.

Smith, Adrian J., R. Eddie Clutton, Elliot Lilley, Kristine E. Aa Hansen, and Trond Brattelid. "PREPARE: Guidelines for Planning Animal Research and Testing." *Laboratory Animals* 52, no. 2 (2018): 135–41. doi: 10.1177/0023677217724823.

Song, Eunhye, Lin Ang, Ji-Yeun Park, Eun-Young Jun, Kyeong Han Kim, Jihee Jun, Sunju Park, and Myeong Soo Lee. "A Scoping Review on Biomedical Journal Peer Review Guides for Reviewers." *PLoS One* 16, no. 5 (2021): e0251440. doi: 10.1371/journal.pone.0251440.

Steeves, Vicky. "Reproducibility Librarianship." *Collaborative Librarianship* 9, no. 2 (2017): 80–89. https://digitalcommons.du.edu/collaborativelibrarianship/vol9/iss2/4.

United States Department of Agriculture. *Animal Welfare Act of 1966* (PL 89-544). Approved 24 August 1966. https://www.nal.usda.gov/awic/animal-welfare-act.

United States Department of Agriculture Animal Care. "Animal Care Resource Guide, Policy #12: Written Narrative for Alternatives to Painful Procedures." 2011.

Chapter Twelve

Expanding Opportunities for Librarians within Institutional Research Activities

Narratives of Engagement

Laureen P. Cantwell, with contributions from
Tracy C. Shields, John Sisson, Nathan Hall,
Emily F. Gorman, Brian Jackson, Daureen Nesdill,
Wendy Highby, Andrea M. Harrow,
Susan M. Harnett, and Megan Sheffield

This chapter provides narratives from librarians working at academic institutions across the United States and Canada who are not necessarily on an institutional review board (IRB) or institutional animal care and use committee (IACUC) at their institution but who are in IRB/IACUC *adjacent* roles at the very least, through the contribution of skills, training, and guidance. These roles engage with research funding and grants discovery and application, scholarly communication, digital preservation, data management, data policy, and governance, research data management, ethics review, instructional design, and data services librarianship. Whether the reader of this chapter is on an IACUC or IRB currently and wants to increase their value to the committee, hopes to be on such committees and seeks relevant skills development pathways, or may not have those circumstances but wants to build strategic, value-added partnerships between their library and these committees at their institution—these narratives offer personal insights into the *skill sets* that critically and usefully connect their roles to the work of IRBs and IACUCs.

The narratives that follow discuss a wealth of important factors, abilities, and soft skills, including but not limited to the value of eagerness and interest, the roles academic librarians play as liaisons and individuals with strong connections to the world of research, the capitalization of our careers with our individuality and uniqueness, the prior experiences we all have, and more. Additionally, to a person, the voices here indicate willingness to take on additional skills, to develop the skills of others and to create resources to that end, to serve at the request of an institution but also to advocate for potential ways of serving as well, and to be a consultant or advocate to and for others. Lastly, regardless of whether a librarian has a teaching role or background, the narratives indicate that the ability to take on and own an instructional role is key. Similarly, and likely due to academic librarians often having to discuss the complexities of (re)search with laypeople, several instances throughout this book and in this narrative highlight the position librarians can take on as middleman between the researcher and the layperson (or study participant), as a means of also protecting the participants (and the researcher/institution). Lastly, librarians are often fairly well versed (and invested) in ethics, in terms of both behavior and the understanding of its limitations, as well as the realities of addressing information reuse. All of these narratives combine to not only make a picture of the librarian as an appropriate IRB and IACUC member, but they also share the unique, tailored, and sought-after skills and activities these librarians have.

SERVING ON THE IRB AND IACUC

Soon after joining library services as the reference medical librarian at Naval Medical Center Portsmouth (NMCP) in 2014, I was asked to serve as a non-scientist, voting member on one of the two institutional review boards (IRBs) there. A strong relationship between librarians and the IRB was already established at NMCP; my colleague has served on one IRB since 2013, and previous occupants of my current position had served as well. Two years later I was invited to join the Institutional Animal Care and Use Committee (IACUC), expanding my role with the IACUC beyond literature search support. I have served on both ethics committees since November 2016.

Similarities and Differences

The IRB upholds the Belmont Report's three core principles: Respect for Persons, Beneficence, and Justice; the IACUC complies with the 3Rs: *replacement*, *reduction*, and *refinement*. Now that my duties overlap both committees, I find that my approaches to these principles overlap as well. My experiences from IRB inform IACUC and vice versa. I may consider "justice" along with the 3Rs for an animal protocol. I may think about "replacement" during an IRB review and ask if research must be done in a vulnerable population.

 I thought I knew what to expect as a new IACUC member based on my experience on the IRB. There are similarities to be sure, but there are also some stark differences. It became clear early on that there are many more regulations and expectations surrounding animal research compared to human subject research, and that those protections are essential.

IACUC requires more time and involvement of members overall than IRB, in large part because of the added regulations. Having searched the biomedical literature for clinicians, I am much more familiar with human-based research, study designs, and protocols. I often need to do more background reading and learning for animal-related research to even understand the protocols so I can make an informed decision as an IACUC member. My duties as an IACUC member include semi-annual inspections along with more continuing education to keep abreast of trends in animal research, so I spend more time on that than I might with human subject research.

Military-Specific Concerns

One aspect of my IRB and IACUC duties that may be a bit different from others is related to my place of work. Most NMCP staff are U.S. Navy active-duty military personnel, and the patients they care for are other service members and their families. Civilian providers and support staff augment active-duty military personnel and provide continuity during deployments.

The human subjects research that takes place at NMCP may include active-duty military as subjects, and animal research is done to support the overall mission. Those factors add another layer of responsibility and need for protection; military personnel could be considered a vulnerable population, as rank and duties may impact how or where research is done, how subjects are recruited, or require added protections for privacy concerns.

IRB members—most of whom are active-duty military themselves, along with civilian counterparts like me—often need to consider factors such as culture, mission critical needs, and other military-related concerns as we look at protocols. The principles found in the Belmont Report and 3Rs offer guidance but cannot speak to nuances that may be specific to a military population.

Some questions I may ask as I review an IRB or IACUC protocol: Can a study based on an aircraft carrier recruit participants without undue pressure? Will involvement in research affect the promotion of the subject or impact mission critical duties? Is this the right animal model for this type of research that is critical to the military? Will the use of these animals in this research lead to lifesaving changes in medicine?

The Research Lifecycle

As a librarian and member of IRB and IACUC, I get to see the "before and after" of research. Like other reference librarians in a hospital or academic setting, I handle the requests for literature searching and training for clinical support, graduate medical education, and research. I provide the necessary information, search strategies, and other help to requesters, but often have no idea what is done with it after I have provided aid.

Being on both committees allows me to see how those requests are incorporated into the research process and the impact they have on guiding research decisions. I have used that experience to expand my knowledge, skills, and abilities with searching in general. I am better informed in how I present searching and results to users, how I handle requests, what questions I ask as part of the reference interview, and the databases I search on behalf of researchers.

Learning more about the research process—human subjects or animals—has also led to changes in training and services the library provides. After seeing questions about citing and proper usage of questionnaires in research, we developed an information session on citations, permissions, and copyright for staff. We have since incorporated parts of that talk into our standard training for graduate medical education and expanded our involvement in "Research 101" initiatives at NMCP.

Literature Searching

From a reference librarian's perspective, the literature search requirements for IACUC protocols compared to IRB protocols are probably the most obvious differences between the two committees. In many cases, the literature search and subsequent review will greatly inform researchers about what kind of research can and should be done. IACUC regulations require searches for duplication and alternatives, with guidance in place to direct where and how to search.

Unlike IACUC searches, there is little regulatory guidance beyond "do a comprehensive literature review in more than one database" to inform IRB literature searching—a gap that has had severe negative consequences for human subject research and IRBs in the past. (For a prominent example, see the Johns Hopkins hexamethonium asthma study from 2001 and the death of a healthy human volunteer.)

Librarian as Expert

Both groups look to me to provide the subject matter expert opinion for anything related to literature searching—where and how researchers searched, what they may have missed and why, and what guidance can be offered to the committee on how concerns might be addressed. I often do "quick and dirty" searches during meetings when other members raise questions about literature reviews, protocols, or other studies.

I have conducted continuing education training for fellow IRB and IACUC members so that they can better "read" and critique searches that may be documented in protocols. When I first joined, I might be the only IACUC member to raise concerns about terms used in an alternatives search or the duplicates search being too focused; now other members readily speak up and point out search-related issues as well and seek my confirmation on their comments. The IRB and IACUC administrators have also suggested to researchers to meet with me for assistance with protocols before they get to the boards or committees.

My perspective is often solicited to evaluate the informed consent forms for human subject research. I and my librarian colleague on the other IRB have used our library backgrounds and consumer health interests to evaluate forms and other study documentation for readability and health literacy levels.

Librarians as Hub and Library Outreach

These approaches have widened the impact of the library as a whole and elevated the library's profile beyond clinical, educational, and research support. The library and librarians are partners in research throughout the process— from finding gaps in the literature and protocol development all the way to finalizing research and publishing.

The intersection of reference work and ethic committees has also led me to connect potential researchers to other interested parties. For example, when a resident requests a literature search from me for a potential project, I may be aware of a similar or overlapping study already in the works, perhaps in a different department. Because of my IRB and IACUC memberships, I am aware of those projects and can suggest the resident reach out to the research group. Because of my connections through IRB and IACUC, I have also had the situation where I have helped make unexpected professional connections through the "I know somebody" approach.

Why I Am There

As I see it, my role on the IRB and IACUC is to provide an "outsider" perspective for researchers and regulatory authorities. Being the nonscientist means I am sometimes in an uncomfortable position, and I must think beyond my temporary discomfort to why I am there. I may need to speak up when I may want to remain quiet, to question someone with years of research experience when I am not a researcher, and to admit to a lack of knowledge and understanding of complex issues when my job is all about information and being informed.

Others in this book have noted the important role of the nonscientist member on these committees, but it bears repeating: on the IRB, I speak for the patients; on the IACUC, I speak for the animals. That is a position that I take seriously and respect for the awesome responsibility behind it.

Tracy C. Shields has a BS in biology from Vanderbilt University and an MSIS in information science from the University of Tennessee–Knoxville. Tracy is the reference medical librarian at the Naval Medical Center Portsmouth.

From Shields's discussion of her role on both the IRB and the IACUC within a military-associated medical setting—and the various responsibilities, perspectives, and skill sets therein—the next narrative zeroes in on one librarian's role on the IACUC and how he developed, and now leverages, his grant research skills. These two narratives—really all the narratives within this chapter—truly highlight the connections between their activities and the title of this book, *Finding Your Seat at the Table*. What's more, these narratives explore the means by which librarians can *grow* in their involvement with IRBs and IACUCs. What comes first—the committee role or the tangential service(s) to the committee? For some, like John Sisson, the IRB and/or IACUC role may open the gates to showcase the specialized skills garnered throughout their education and their career.

IACUC PARTICIPATION
AND RESEARCH GRANTS DISCOVERY SKILLS

I have been a member of the UCI (University of California, Irvine) Institutional Animal Care and Use Committee (IACUC) since 2004. I became a member when a previous librarian was retiring (as the nonscientific member) and asked if I would be interested. As the research librarian for biology since 1990, I was eager for the chance to learn more about my faculty's research interests. UCI has a large School of Biological Sciences as well as a College of Medicine. There is a great deal of animal use on campus and a vibrant research program. The UCI overall annual research budget in 2020 was $529 million in grants and contracts.

Librarians tend to operate on most campuses in the written research world, helping faculty and students with submitting research papers, subscribing to research journals, and helping the faculty and students access these sources. The research grant cycle and the difficulties associated with funding the research is rarely part of what most subject librarians regard as part of their faculty liaison. This "hidden" work is not supported directly by the library. As part of a small number of librarians working with these questions, I have created guides to research grant resources and orient graduate students about the grant application process.

My ability in assisting with grants is a combination of experiences as a graduate student in biology and opportunities and learning through my job. In graduate school I served on a graduate grant review board for two years. As part of the board, I would read 150 applications per year to help award five grants. I also worked in a laboratory that had several research grants. Conversations with senior students in the laboratory and my graduate advisor gave a sense of how research funding deadlines can drive research efforts. I would be asked to supply short statements of my work that my advisor could adapt for grant reporting and applications.

As a research librarian I interview faculty about their research, and they may forward copies of their applications or general summaries as preparation. I was also asked by my university librarian to explore how the library could support grant writers. I explored various institutional research funding resources and institutional research productivity tools that the library was requested to buy. I created a summary table and recommendations comparing which tools we already owned and which would be useful to add. I also participated several times as a member and chair of our librarian research grant review committee. During my two 3-year terms, we assisted applicants and reviewed grants for recommendations to state-wide funding committees.

Very few faculty and graduate students approach me directly for help with their applications. However, when they do inquire, I can refer them to resources and recommend search strategies since I am familiar with the application process. I think this increases their perception of the librarian as a peer in the research enterprise beyond simply finding articles. I also have had opportunities to work with the Office of Research staff who assist faculty. They have helped me promote campus funded resources in my research guide and the ongoing relationship helps with creating a span of library-funded and campus-funded resources. I have been invited to training sessions and to discussions about the place of institutional research productivity tools in the campus research ecosystem.

These experiences feed into contributions to discussions during IACUC meetings. As the campus continues to explore different outside funding opportunities, my background knowledge helps me to understand the role IACUC-approved applications and IACUC supervision play in meeting the assurance agreement of new sources of external entrepreneurial funding and animal testing.

A key part of the IACUC application process is the "brief non-technical, lay-language summary of your project." This short summary is one of the few pieces of the application that is treated like a public record. It may be used by an organization to prepare press releases or discuss the project with the media. Many federal applications have similar requirements as part of their project summary. An important task on the IACUC committee is reviewing and assessing these sections.

My knowledge of the research grant process and its needs is helpful in giving applicants constructive feedback. My experiences with helping undergraduates writing in the sciences has been invaluable in helping researchers who must rewrite and clarify this section. I have worked with undergraduates to learn how to write abstracts of their research project. I point faculty to the same basic idea that abstracts should be understandable by someone who is not in the discipline or even accessible to a nonscientific audience.

> *John Sisson has a BA in general biology from the University of California, San Diego; an MS in zoology from the University of Maine; and an MLIS in library science from the University of Hawaii. He is currently the biological sciences librarian, University of California–Irvine Libraries.*

Grant discovery skills provide a useful bridge from the searching, observational, and deductive reasoning skills librarians must employ with other databases and research topics and into the grant writing skills addressed by Nathan Hall below. Hall also aids with data management, retention, and storage concerns as well as topics within research reproducibility, speaking about the challenges and benefits of having these roles and participating on his institution's IRB. In that role, he took on a new responsibility for any librarian at Virginia Tech and reflects on the time, effort, and professional development required, but he ultimately recognizes the preparation that role has given him in crafting his own original research projects and the capacity for making a broader impact on the field of librarianship (e.g., at the consortial and/or national levels).

DATA MANAGEMENT, GRANT WRITING, AND RESEARCH REPRODUCIBILITY: BROAD BENEFITS

I am an associate professor in the Virginia Tech University Libraries and I am possibly the first librarian to serve on Virginia Tech's institutional review board (IRB). This appointment gives me access to additional training and knowledge that enables me to advise my library colleagues on human subjects research, and it also gives the library a seat at the table of an important component of the university's research enterprise. The role has required extensive training, but it has also served my own research through better understanding of research methods and ethics. Several other service opportunities and research endeavors prepared me for my appointment to the IRB. While there are certainly different avenues that a librarian could take to prepare themselves for that role, the rest of this narrative will cover the experiences that I have drawn upon as an IRB member.

A number of projects and opportunities prepared me for this work. A significant part of my role at Virginia Tech, since starting in 2011, has been in scholarly communication, digital preservation, data management, data policy, and governance. My dean asked me in 2012 to begin supporting researchers with data management planning, and to promote the University Libraries as a partner for data deposit and curation. During that year, I wrote around thirty data management plans, and I also helped write some sample language for the Association of Southeastern Research Libraries (aserl.org) and the Southeastern Universities Research Association (http://www6.sura.org/) for university level policies on data management. I later served on Virginia Tech's Commission on Research, where I helped review and revise university policies including the Human Subjects Research Policy; the Policy on Ownership and Control of Research Results; and the Animal Research Policy, among others. I spent six years as a member and eventual chair of the Association of College and Research Libraries Research and Scholarly Environment Committee. This role required thorough knowledge of federal data policy and the position drafts positions for the association's responses to federal requests for information (RFIs) on research policy. Finally, my PhD in information science from the University of North Texas helped prepare me through coursework and experience in research methods and through managing a human subjects research project. Together, this work led to greater understanding about how data management supports transparency in review and publication, and in reproducibility and reliability of scientific findings.

Grant writing and administration has been a major thrust for me for the past four years. I participated in Virginia Tech's Proposal Development Institute, which provided lectures and mentoring on different grant development and grant management topics, such as budget design, personnel costs, overhead, direct and indirect costs, automatic escalation, research policies, post-award compliance, broader impacts, and how to write a successful narrative. Since then, I have been the principal investigator (PI) on three externally funded projects, including Developing Library Strategy for 3D and Virtual Reality Collection Development and Reuse (https://www.imls .gov/sites/default/files/grants/lg-73-17-0141-17/proposals/lg-73-17-0141-17 -full-proposal-documents.pdf), Community Development Model for Digital Community Archives (https://www.imls.gov/sites/default/fils/grants/lg-15 -19-0137-19/proposals/lg-15-19-0137-full-proposal.pdf), and Entomo-3D: Digitizing Virginia Tech's Insect Collection (https://www.clir.org/hiddencollections/funded-projects/). I am also a reviewer for Council on Library and Information Resources' Digitizing Hidden Collections program. While proposals to that program don't usually include human subjects research, they do require review of legal and ethical considerations and privacy concerns.

I volunteered for Virginia Tech's IRB two years ago, offering my knowledge of data management as an asset, and this happened to be an area that they wanted to add within the next few months. Our Human Research Protection Program (HRPP) includes IRB members, as well as staff (one director, one administrative specialist, and five protocol coordinators). Our HRPP was understaffed at this time, leading to backlogs, which were frustrating for faculty. When I volunteered to join the IRB, I was able to participate in the hiring process to fill the empty HRPP positions by meeting with some of the applicants. Through these meetings with candidates from a variety of institutions, I learned how much IRBs differ in composition and in focus by institution. Virginia Tech for example, has a lot of engineering studies and driver studies, due to the Virginia Tech Transportation Institute. My joining the IRB also coincided with the Virginia Tech Carilion School of Medicine merger with Virginia Tech, thus increasing the volume of clinical research in the IRB. I have had to learn a lot about clinical trials in a short time in order to better understand the proposals I was evaluating.

Most of my first year on the committee was spent gaining basic competence, and I underestimated the time I would spend on training. Through our institutional subscription to CITI Program (www.citiprogram.org), I was required to take a course for IRB members, as well as Basic Responsible Conduct of Research; Biomedical Research; and Social and Behavioral Research. Each of these courses included ten to twenty modules, with a test for each module. Combined, all of this instruction took more than fifty hours over two years.

While I can bring my Open Knowledge values to this work, I also have to understand the ethical limitations to open knowledge, and the ways that members of marginalized communities have been exploited. IRBs do not care about reproducibility or data sharing for the sake of openness. They care about protecting human subjects, and they care about research validity (some institutions have separate review committees for this). A lot of research would be unethical to make public, for a variety of reasons. A significant aspect of human subjects protection is protecting their privacy and their confidentiality, which requires some understanding of data security. Researchers often say in their protocols "data will be on a password-protected computer." I consider this language to be inadequate, because a password on a personal computer could be "123456" or "password." Or it could be on a sticky note stuck on a laptop. I usually ask for a change to something like "data will be stored on a university-secured network," which at Virginia Tech means a sufficiently complicated password, dual authentication, and in most cases, data encryption, but doesn't burden the PI with those details. While I can bring my Open Knowledge values to this work, I also have to understand the ethical limitations to open knowledge, and the ways that members of marginalized communities have been exploited.

In summary, a large part of my work in scholarly communications and digital preservation has been in managing information for reuse, working with contracts, policies, governance, technology management, and risk mitigation. These led to work on other aspects of the research enterprise, which in turn prepared me for service on the IRB. I have been lucky to have opportunities in many of these areas, but candidly, there has been a significant learning curve in my becoming a fully contributing IRB member. My own data collection has been expedited or waived, and not having experience in research that involves significant physical, social, psychological, or economic risk, nor experience working with populations with reduced autonomy, has been a limitation. I have had to remind myself what I bring to IRB that is unique and meaningful. While many IRBs and IACUCs may at first be hesitant to include a librarian, persistence pays off, along with understanding their needs and how to meet them.

Nathan Hall has a BA in English and music from St. Lawrence University; an MLS in digital image management from the University of North Texas; and a PhD in information science, University of North Texas. He is the director of digital imaging and preservation services at Virginia Tech University Libraries.

Hall's narrative connects us with his background and how he leveraged that into a volunteer role with his institution's IRB. This provides an easy arc into our next narrative, from Emily F. Gorman. Gorman writes from her role as a nonscientist member of an institutional review board. She shares her motives for initially getting involved, several years ago, which includes the value of IRB participation as a networking tool while she was new to her institution. Gorman also details her "fit" within the work of the committee, most especially the benefits and value of the nonscientist lens and voice—sometimes the *only* nonscientist present to *be* a voice—within an IRB. She makes a strong case for that role's importance, while also acknowledging needing to get a "feel" for the work itself, most especially the act of serving as a proposal reviewer and within the context of informed consent documentation.

DEFENDER OF THE PEOPLE: REFLECTIONS
OF A NONSCIENTIST MEMBER OF THE IRB

I have served on my university's institutional review board (IRB) as a non-scientist member for three years now, and it has been an illuminating experience. One of my original motivations for becoming involved was that it was essentially the only opportunity available to me as an early-career faculty librarian for joining a university-wide committee. Additionally, I had heard from colleagues in my department who were also on the IRB that it was an interesting way to learn about the research occurring at our institution and to meet faculty in other schools. So about six months into my new job, I took the plunge.

Joining the IRB was fairly straightforward. I reached out to our human research protections office (HRPO) to express my interest, and after providing my curriculum vitae (CV), I received a nomination letter from the university's chief academic research officer offering me a three-year appointment to the board. Then the logistics of completing all the required training began. Several online trainings covered the history of research ethics and IRBs, how to appropriately conduct human subjects research, and the responsibilities of IRB members. Once I completed all the online modules, I attended an in-person orientation session and observed portions of two different IRB meetings before being assigned to my first meeting as a full board member.

Attending an IRB meeting involves the time spent in the actual meeting itself as well as the time spent before the meeting completing reviews of the research protocols. As a nonscientist member, I typically serve as the second or third reviewer on any new research protocols, so my comments are supplementing those of one to two scientist reviewers. If a meeting is only reviewing modifications to, or annual renewals of, approved studies, I am usually not assigned to review anything and have to attend the meeting simply to vote.

Completing my first review of a new research protocol was a daunting task, made easier by consulting my library colleagues already on the IRB for guidance. I was unsure about how much detail regarding the study background and procedures to include in my write-up, particularly if I didn't fully grasp the science. I also didn't know how strict or thorough I should be—was it necessary to correct every typo, even if they were minor and didn't affect understanding of the content? How demanding should I be in terms of plain language explanations of all the jargon in the informed consent documents (ICDs)? The answers to these questions all boiled down to one guiding principle—does it affect participant safety? If a typo is minor and not endangering participants, it's likely not necessary to burden the research team with making those modifications (which sounds as if it would be simple but, in reality, it involves several steps within the IRB protocol management system that would delay the start of the research). However, if a lack of plain language in the ICD means a participant wouldn't fully understand the study procedures or risks, that needs to be addressed. As I attended more meetings and completed more reviews, I grew increasingly comfortable making this distinction.

One of the most intimidating things about being a nonscientist member of the IRB is that I am typically the only one at any given board meeting. More often than not, the room contains medical doctors, PhDs, and me. At my first few meetings I noticed that for a number of the protocols I had reviewed, I was the only one expressing concerns (usually about ICD language) that would warrant deferral of the studies. I then worried that the scientists would regard me as a nuisance standing in the way of approving studies. Fortunately, that anxiety was unwarranted, as IRB administrators and other board members conveyed their support and gratitude for my work, and over time I grew more confident in my role.

The whole point of having a nonscientist serve on the board is to provide a unique perspective, so it makes sense that I would be identifying problems that scientific experts would not notice, such as too much jargon in the consent forms. However, it's one thing to know that you are supposed to be providing a different viewpoint and another to actually have to defend your lone opinion to a group of experts. What has helped me is not only the encouragement of fellow IRB board members who appreciate my work but also telling myself that when it comes to participant understanding of research, I *am* the expert. As cheesy as it sounds, the nonscientist really is the defender of the people when it comes to IRB review. I can read the protocol, especially the consent documents, from the perspective of a potential study participant. If I don't understand something, other people thinking about enrolling in the study will probably be confused too; and if I don't mention it in my review, that problem likely won't be addressed by anyone else. I quickly came to realize that the other board members respect this role and appreciate having someone making sure they don't overlook issues that will affect participant understanding. The feeling that I didn't belong in the IRB meetings lasted only a short while.

Although it occasionally can be frustrating and extremely time-consuming, I have never regretted my decision to join the IRB. It adds a welcome variety to my work and has enabled me to meet new people from all over campus. And when I need to vent about the meeting that ran late by almost an hour or about reviewing a forty-page consent form, I can commiserate with my library colleagues who are also on the board. It is rewarding to know that what I am doing makes a difference to hundreds, if not thousands, of people enrolling in research studies at UMB. Here's to the next three years of service!

Emily F. Gorman has a BA in psychology and religious studies from the University of Virginia; and an MLIS with academic libraries specialization from the University of Pittsburgh. She is the research education and outreach librarian for the School of Pharmacy, Health Sciences and Human Services Library, University of Maryland.

While Gorman's narrative highlights the unique, beneficial contributions she is able to bring—literally—to the IRB table at her institution, Brian Jackson brings his contributions as a data librarian to the forefront. Jackson, at Mount Royal University, is located in the traditional territories of the Niitsitapi (Blackfoot) and the people of the Treaty 7 region in Southern Alberta, Canada. Jackson's additional comments, as a data librarian and a member of the Human Research Ethics Board, highlight the benefits of being recruited for campus service opportunities based on his expertise, which extends into being able to serve both the committee, as an insider, and the campus community at large. Within his committee participation, Jackson rightfully highlights his capacity to aid researchers in anticipating and addressing recurring and burgeoning concerns. These concerns include data collection and/or storage beyond one's national borders—which increases in likelihood as multi-national, multi-institutional collaborations grow—versus the specifications of local/provincial oversight; discussions of the provenance or ownership of data, and how and where that data is stored—again, especially in multi-national contexts, but also in terms of commercial data interests, and how these concerns interact with sharing information storage, ownership, and access details within participant consent documentation. Jackson's positioning for early intervention in research plans and decisions has impacts not just for the committee and the campus community but clearly for the welfare and protection of research participants as well.

RESEARCH DATA MANAGEMENT
AND ETHICS REVIEW PARTICIPATION

I am a data librarian at Mount Royal University, an undergraduate university in Calgary, Canada. In 2017, I was recruited to the university's human research ethics board as a data management specialist. In that position, I provide guidance to members of the board, as well as to the wider institutional community, on ethical practices for research data collection, management, and publishing.

Familiarity with the policy structures that govern the handling of research data may be the most crucial skill for this work. In Canada, institutional research ethics policy development is led jointly by three federal research agencies and documented in the *Tri-Council Policy Statement: Ethical Conduct for Research Involving Humans* (TCPS2). As universities in Canada are largely subject to provincial oversight, the management of research data may also be regulated by provincial privacy legislation, the scope of which varies from province to province. Going further, internal policy at my institution governs some aspects of research data management, including storage and terms of retention. Part of my role is to interpret and provide information about the policy environment with regard to the fitness of research data practices. I do so generally by developing web-based guidelines on research data collection, storage, and archiving for the benefit of the research community, and specifically by providing commentary on the data practices within applications during the ethics review process.

A recurring concern among researchers in Canada stems from the use of data collection and storage tools that are based outside of the country. Depending on the nature and location of the research, research data may be required by law to be stored within Canada or even within a particular province. Regardless of the legislation, the possibility that personal information collected through research may be subject to access laws outside of Canada requires informed consent of participants. Part of my role is to investigate how and where data are stored by software companies to help researchers anticipate the provenance of the data and inform participants. For this, a baseline understanding of how internet companies store, transmit, and encrypt data is beneficial, but in an environment of evolving tools and research methods, providing this information to researchers is a process that requires continuous development.

Like many libraries, mine has developed a suite of open research data programs and services, including data archiving through our institutional data repository. My involvement in the ethics review stage affords me the opportunity to provide researchers with information about open data and the process of obtaining consent for data archiving, and to identify any open data requirements of research funders or publishers that might apply to the research. This early-stage intervention benefits the research enterprise in terms of funding compliance and the publication of additional scholarly contributions, while serving to connect researchers with library data services and promote open data. Broadly, my position has created awareness about the expertise and supports that exist in the library that benefit all stages of the data management lifecycle.

Brian Jackson has a BA in history from the University of Guelph and an MLIS in librarianship from the University of Alberta. He is a data librarian and subject specialist for earth and environmental sciences, policy studies, political science, and economics at Mount Royal University.

Perhaps surprisingly, not all "data librarianship" positions are the same. However, as seen in other narratives here, Daureen Nesdill at the University of Utah emphasizes the impact that a librarian's subject matter background and prior work experience can have upon involvement with IACUC and IRB governing bodies. While her master's degree in librarianship qualifies her for positions within academic libraries, her academic background in biology, zoology, and laboratory animal science not only made her competitive for a science librarianship position at the University of Utah but also for roles within the campus IACUC and for meaningful connections with U of U's Office of the Vice President for Research (OVPR). In particular, Nesdill's work through the OVPR, rather than through the IACUC or library, has brought out opportunities for meaningful outreach to stakeholders, especially through workshop creation and delivery, pilot studies of research tools (like electronic lab notebooks, or ELNs), leading and participating in information sessions, and writing blog posts. Her outreach has also appropriately included services promotion as well—for example, assistance with literature searches, data management plans, data discovery, university and federal policy and more. While her titles do not indicate a role as an "outreach librarian" in the way that the latter is typically meant in library job jargon these days, it is clear that outreach is a valuable, and valued, component of her work as research data management librarian—as it should be.

LABARCHIVES, RESEARCH DATA MANAGEMENT, AND MULTI-PRONGED OUTREACH EFFORTS

My undergraduate degree in biology and laboratory animal science led to a position as animal technologist in an endocrinology research laboratory. After five years my allergies to the animals led me to further my education in ecology; but, before obtaining the PhD, I switched to library science. When I landed a position as science librarian at the University of Utah, I volunteered to serve on the IACUC. Because of my undergraduate degree and experience in both bench and field work I have been "volunteered" by the vice president for research to be a member of IACUC for the past sixteen years. I even chaired the committee for two years. Fortuitously, when I took on the role as the data management librarian, I already had a relationship with the people in the Office of the VP for Research (OVPR). This assisted me with promoting my new services to researchers campus-wide. The OVPR has a weekly blog allowing me to promote information concerning data management. With the many changes happening in this new field, I am kept busy contributing to the blog. I also promote my services whenever asked to assist with the required

IACUC literature searches. I inform the researcher of my role as the data management librarian (e.g., writing data management plans, locating data and data repositories, campus resources, university and federal policies). Some of these sessions have led to invitations to present at departmental meetings. I now also teach for the undergraduate research programs and as part of the ethics course required for all graduate students.

I am able to teach workshops in data management through the OVPR's research education program. This program teaches researchers all aspects of applying for and managing grants, safety issues, ethics, policies, and so on. The OVPR sets up the classes and provides treats. I just have to show up and speak. In my workshop I include campus resources such as the IACUC. I explain the procedures for applying and how the different environmental safety offices work with the IACUC so that all regulations and policies are met.

About 2009, electronic lab notebooks (ELNs) became available for academia. Researchers were working in a hybrid world—part electronic and part analog. ELNs would facilitate the management of the research and address the growing concern with reproducibility. I started an investigation of the many ELNs available and discussed the issue with the associate VP, who supported me in my efforts. We surveyed the campus, found interest, and then initiated a yearlong pilot study using LabArchives to determine the issues with implementing an ELN campus-wide. Toward the end of the year, an information session was held with the pilot participants providing their experiences to attendees. This led to a campus site license paid by the OVPR and my added role as campus administrator for LabArchives. It is time-consuming for a research group to become comfortable using LabArchives, but through assisting IACUC members, holding workshops, and writing blog posts, we now have over one thousand users of LabArchives on campus.

Daureen Nesdill has a BS in laboratory animal science from UNY Downstate College of Health Related Professions; an MS in zoology (and ABD in behavioral ecology) from Auburn University; and an MLIS from the School of Information Science at the University of Alabama. She is the administrator for LabArchives (at the university level), research data management librarian, and full librarian at University of Utah.

Within the insider club of academic librarianship, a librarian's fluency and expertise in search skills should come as no surprise and yet, as with Nesdill's narrative, these skills are strong within the librarian population. As part of writing her chapter on the history of IRBs, Wendy Highby delved into and explored her institution's university archives for a sense of *their* history. This endeavor proves to be useful and an engaging way to place the history of her institution's IRB within the overall historical contents of institutional review boards. Not only is this an interesting way to connect librarians and IRBs/ IACUCs with their institutional archives, and the overall timeline of IRB history, but it also shows how this work makes librarians even more valuable assets to their institution.

CAPTURING THE HISTORY OF YOUR LOCAL IRB, WHY IT NEEDS YOU, AND HOW TO HELP

Archivists have the means to capture and preserve the history of their local IRBs. Finding aids can highlight information excavated from manuals and minutes. That data can tell the tale of the local IRB's founding and development. Primary and secondary sources of institutional history can be highly instructive for those who follow. Who held leadership positions on the IRB? How did they help shape the research agenda of the organization? Just as archivists have an important role in curating this history, so do subject librarians. Those who liaise with faculty and other researchers in various programs and departments have insight into the research interests and career trajectories of those scholars. Subject librarians are ideally situated to witness the research activity at their institution. They hear the good news when grants are awarded and may know of the nature of planned or recently conducted studies.

I contacted the archivists at my institution and discovered that the early history of our IRB could be found among our graduate school records in the form of a "Theses and Dissertations Manual" and an "Annual Report of the Graduate School." The latter noted the approval of the "University-wide guidelines for human subject research." Once you know the history of your local IRB, you can insert the significant dates and events (such as formation, leadership changes, etc.) into your customized timeline (building on the timelines discussed in chapter 2 of this book). The 1982 origin date at my institution becomes more meaningful in context. This was toward the end of the era in which IRBs proliferated. That the University of Northern Colorado (UNC) would be a relatively late adopter makes sense, given that it has no medical school and its historical role as Colorado's normal school. Where in the timeline would you place your organization?

Role-Related Knowledge and Discipline-Specific Issues

Knowledge of the political history of the IRB at one's institution is essential for those who serve in leadership positions. And for those who serve as reviewers, though less essential, it is still of great help. Reviewers may keep their heads down and engage in the relatively solitary task of reviewing the IRB submissions delegated to them, correctly focusing on the rights of the participants. Yet knowing the context can help you to correctly interpret reactions of investigators. For instance, reviewers may be blindsided by pushback they receive from investigators in the social sciences if they are unaware that historically, representatives from the social sciences were often excluded from the regulation deliberation processes. Many colleagues have confided to me that they feel intimidated or anxious about the prospect and process of IRB scrutiny.

If you are a subject librarian, you need to know the discipline-specific aspects of IRB history. For example, I, as a liaison to the psychology department should be aware that the American Psychological Association (APA) developed a code of ethics in 1952–1953 that included a consent requirement. I should know that Dr. Irwin Berg championed informed consent in a 1954 *American Psychologist* article. I also should be cognizant of the issues that Milgram's research raised in the 1960s. And finally, I should be familiar with the controversies surrounding psychologists' involvement in the research participation and rendition of prisoners of war. Knowing the political and historical context of the local IRB (as well as the federal- and international-level contexts), and understanding discipline-specific histories, makes one a better subject or reference librarian. Liaison librarians can incorporate and integrate all they know when they advise their library users. As research ethics- and IRB-related questions arise, they will have more nuanced, accurate answers and can make better referrals.

What can you do to influence the trajectory of this ever-expanding system of IRBs? Can you be the change you wish to see? Do your part. Start with preserving your local history, and remember to insert yourself into the timeline, supporting and nurturing what you value and working to change the things that you can. Are you an overt opposer, an apparent accommodator, or an active engager? (Labels for attitudes toward IRBs identified by Judith Taylor and Matthew Patterson.) Be self-reflexive and then act upon your self-analysis. If you honestly oppose the IRB system, work to create an IRB with which you can productively and proudly engage. If you want to change things on the federal level, work through advocacy groups and learned societies that employ lobbyists to speak for you. Examples of position papers from advocacy groups such as the American Association of University Professors (AAUP) and the American Educational Research Association (AERA) are cited in the bibliography of chapter 2.

Lastly, I'm including some continuing education suggestions. The OHRP website (Office for Human Research Protections) offers several features that will help the information professional to keep up to date on changes to IRB regulations. Sign up for the OHRP announcements listserv at this URL: https://

www.hhs.gov/ohrp/news/sign-up-for-announcements/index.html. Avail yourself of their educational resources at this URL: https://www.hhs.gov/ohrp/
education-and-outreach/index.html. Read the *IRB Advisor*, *IRB Ethics and
Human Research*, and *Journal of Empirical Research on Human Research
Ethics*. If you haven't already, get to know the people in the office of sponsored research at your organization. If you are not directly involved with the
IRB at your workplace, introduce yourself. Your local IRB needs you.

> *Wendy Highby has an AAS in Paralegal Studies from the Community College
> of Denver; a BA in sociology from the University of Colorado at Denver; and
> an MLS from Emporia State University. She is currently a professor and social
> sciences librarian at the James A. Michener Library of the University of North-
> ern Colorado.*

Highby's unique and inspirational lens on finding a seat at the IRB
"table" is replete with a sense of expectations and professional development
resources to help librarians not just be *at* the table, but to be a useful, beneficial, and informed *participant*. Andrea M. Harrow, the author of the next
narrative, provides a distinctly boots-on-the-ground description of medical librarianship in a hospital setting, which includes facilitating research reporting
activities. This narrative—which also involves COVID-era realities, hospital
acquisitions, and other compelling variables—details a kind of day-in-the-life
of a hospital librarian's role with her research oversight committee.

HOSPITAL RESEARCH OVERSIGHT

For the past ten years, I have been the administrative coordinator for hospital-
based research reporting to the medical staff management department. Every
morning I stop in to chat with my department colleagues to catch up on the
latest hospital and personal news before I make my way to the medical library. At my desk with a cup of tea, I wake up my computer and open my
emails to review the latest remarks in three ongoing hospital-research-related
email conversations. One of our physician investigators in the neonatal intensive care unit, or NICU, is trying to arrange for a site visit with a research
sponsor. He does not have an on-staff research coordinator and does the data
collection and reporting work himself or persuades hospital staff involved
in the research process to help. I am delighted that the clinical trials nurse
in oncology, who is also cochair of our enterprise-wide research oversight
committee (ROC), is a good communicator and manager and has asked the
right questions of the sponsor rep, doctor, and hospital departments. Because
our compliance office and health information department do not allow virtual

screen shares of the patient record, she has offered to collect and print needed documentation to send to the sponsor. In-person site visits are difficult during the global COVID-19 pandemic. Travel is discouraged and social distancing restrictions remain in place.

Our hospital was recently acquired during the last days of 2019 and is now the third hospital in a small Southern California health care system. The change in culture, staff, and processes during this time has been dramatic. I considered myself knowledgeable about hospital-based clinical trials but the last nine months have exposed me to even more of the detailed processes. The ROC functions in a similar way to our previous hospital research advisory committee. We continue to review proposed research protocols and informed consent that when approved by the ROC may then be sent to WIRB or another OHRP-registered IRB for review and approval. During protocol review, I am able to use my library skills to assist with literature searches when there is a patient safety question or retrieve articles cited in the protocol references. The majority of my role with hospital research, however, is with the administrative management as I advise prospective investigators and answer questions about hospital research program policy and processes. Some of my duties in the past included maintaining our Federalwide Assurance (FWA) registration and reinitiating our sam.gov account and I continue to manage our research study records by maintaining the files. I am involved in the review of clinical trial agreements but, in the past, I would pass everything to our hospital legal counsel for approval. Now I review the contract and complete a checklist to ensure necessary language is in place and then submit to our contracts department. In regard to the financial aspects of a clinical trial, in my past role I would hand off the clinical trial research budget to the finance department committee member to review. Now I also work with the researchers to report quarterly enrollment and billing to the committee. Investigators and coordinators must show that research is billed correctly, whether to the sponsor or to the insurer. Risk management and compliance are emphasized, as is coding and billing oversight.

It is gratifying to work in a health care environment that is adaptable enough to accommodate research when a primary investigator (PI) is highly motivated or when a global pandemic emerges. A few months ago, when we were scrambling to find ways to manage the first gravely ill patients with the SARS-CoV-2 virus, the research committee was called on to review experimental treatments for this novel communicable disease. One of our hospitals joined the remdesivir trial, and all three hospitals joined the convalescent plasma trial. All three of our previously employed cardiology clinical trials nurses were on furlough due to the plummeting census as people stayed home and avoided the hospital. I assisted two of our doctors in registering for the trial and an operating room (OR) nurse was recruited to manage the trial since the OR was closed to non-emergent procedures. The oncology trial nurse previously mentioned developed procedures to follow and trained the other nurses in what to do.

It is almost time to leave for the day and another email reply appears on a different thread but involving the same NICU researcher. A different, more complicated NICU study is not moving forward because a hospital-employed study coordinator is required for oversight of the study. We can only bend our rules and processes so much, and sometimes we just have to draw the line and say no. However, I do not think this is the last we will hear about this from our determined research investigator. I have no doubt he will come up with alternate suggestions to try and move this research protocol forward.

Andrea M. Harrow has a BA in religious studies from the University of California–Santa Barbara; and a masters of information services management from the London Metropolitan University. Andrea is the medical librarian at PIH Health Good Samaritan Hospital, Los Angeles.

With Highby's narrative, we spent time in the archives, peeking into the history of an institution's IRB—then in Harrow's narrative, we zoomed to the very real present of hospital librarianship. Hospital acquisitions, COVID restrictions, shifts/expansion of job responsibilities and expectations—all mixed in with handling classic tasks like literature reviews and search assistance. Librarians often enough get questions about whether the Internet (or Google) has made their work obsolete, and yet librarians commonly recognize the reality is almost just the opposite. With soaring information publishing, access, searchability, and discovery methods, and the complexity of the deep web (vs. surface web and dark web), searching may *feel* easier to do but may not in actuality be easier to do *thoroughly* or *successfully* or with a particular *standard* in mind. Not only do librarians have expertise in (re)searching; librarians also highly value and specialize in the act of (re)search *instruction*.

John Sisson's second piece in this chapter addresses just this—instructional design approaches to discovery methods using databases. Interestingly, Sisson notes not just librarianship as the source of these skills but also his role on UC Irvine's IACUC and the specialized training he received there and through connected opportunities (e.g., at UC Davis). And yet, if it is a case of what came first, the librarianship, and his subject matter background are the nexus point for the IACUC participation and the opportunities that arose as a result of that committee participation. Specifically, here, Sisson's opportunities to teach and grow stakeholders' knowledge and skill sets have led to committee encouragement to guide researchers through consultation, particularly on alternatives searching, as well as his creating user guides, templates for searchers, and more—bringing the library into heightened visibility within IACUC documentation and practices.

CIRCLING BACK ON SISSON'S WORK:
SEARCH TECHNIQUES AND FINDING ALTERNATIVES

An essential part of the IACUC research application is the alternatives search. Researchers are asked to explore the scholarly and professional literature for alternatives. These are the 3Rs: reduction, refinement, and replacement.

Faculty/researcher searches need to satisfy the IACUC that they meet the standard. These search techniques are not intuitive, rather they are taught with examples, sample searches, feedback on results, and in some cases consultations. This is a key role for librarians because of our experience in literature database searching but also because of our ability to teach the concepts of database searching. These alternatives concepts are not amenable to keyword searching. Instead you need to create larger search sets and then add terms to narrow them to relevant concepts.

As a librarian my preparation for this specialized user instruction has had two stages. The first was my professional experiences and training. My knowledge of expert database searching and how to teach it came through classes in the use of PubMed, BIOSIS, Web of Science, Current Contents, and so on. Practice and continual use of these databases for reference gave me the confidence to teach and correct my faculty.

The second was specialized learning and training after I became a member of the UC Irvine IACUC. Our IACUC supports new member training in a variety of ways. We had mentoring by faculty members when assigned grants to review. We would also devote regular time during the IACUC meetings to short training discussions. Individually it supported my attendance at an animal use professional conference and invited me to UC Davis Animal Welfare search training. On my own I looked for other colleagues' examples and user guides. These existing pages gave me a framework for how these techniques can be shared and tailored to my campus.

The result was recognition of my search skills by the IACUC committee. I was asked to comment on alternatives searches in the application review meetings. I was encouraged to offer suggestions that could be sent to the faculty member in the alternatives searches that needed revision. I was also offered as a resource that individual faculty could consult with if they were having trouble meeting the IACUC Committee need for an effective alternatives search.

My instruction designs were built on these experiences and feedback I received. I created IACUC resource user guides and search templates for use by the committee. Some of these templates and resources were included as links in the IACUC application instructions as places to go for further help. These tools and resources made the library more visible as part of the IACUC research process. Indirectly working with IACUC, I have had the knowledge to select animal use books and alternatives resources that would supplement the collection. I was also able to highlight books and videos about animal care and anatomy that might be of use to both faculty and undergraduates.

I share with animal researchers and other faculty an understanding of their specialized needs. It creates a specialty within the library to make effective referrals when we receive questions. By having the resources and knowledge, we effectively support our campus's substantial research mission. This institutional support has resulted in commendations from our university librarian about the comments she gets from senior faculty and the deans from the School of Medicine and Biological Sciences, and the vice chancellor of research about the support the library is giving them.

> *John Sisson has a BA in general biology from the University of California, San Diego; an MS in zoology from the University of Maine; and an MLIS in library science from the University of Hawaii. John is the biological sciences librarian at University of California–Irvine Libraries.*

Sisson shares the variety of ways his background was a keen fit for IACUC membership at his institution. The next narrative, from my coeditor Susan M. Harnett, reminds readers that *not* having close subject matter ties shouldn't deter them from interest in, pursuit of, or serving on institutional regulatory boards and committees. That nonscientist role comes back into play here, as an asset when others have a science background that makes them ineligible to fit the committee's needs. Importantly, Harnett emphasizes the IACUC's responsibility to ensure compliance and humane treatment—at the very least, these are useful arguments to keep in mind for sharing with stakeholders, particularly if the reader feels passionately about those variables.

MEDICAL LIBRARIAN, NONSCIENTIST
VOTING MEMBER, IACUC, 2016–CURRENT

I was asked to serve on the IACUC on my second day of a new job at the university. I had some previous experience serving as an ex-officio member at another institution. The librarian who was originally asked could not serve as she had an undergraduate science degree. My undergraduate degree is in film, which makes me the ideal nonscientific member candidate.

Though there are times when I feel my lack of scientific knowledge is an impediment to my service, I understand that that isn't my job on the IACUC. One of my fellow board members stopped me after a meeting and asked, "How are you finding being on the IACUC?" My response was that when I first started on the board, I understood about 1 percent of what was going on, and after three years, I was up to about 5 percent. He laughed and said, no one understands the science if it isn't their area. The oncologists don't understand the nephrologists. The virologists don't understand the microbiologists. No one understands the biostatisticians. But we're not only here to make sure the scientific premise is sound; we're also here to make sure the protocol follows the law, and the animals are treated humanely.

Service on the IACUC has taught me a lot about research and animal models. Like most of the people on my IACUC, I am a pet owner and animal lover. I would not have chosen a career in which I did animal research, but I am able to understand the necessity of using animal models. My role is not only to ensure compliance and humane treatment of the animals, but in many ways it's to give voice to those animals. If I hear something in a meeting I can speak up. If I review the literature search and it's not done properly, I can ask for revisions. If I spot a deficiency during a site inspection, I can have it corrected. If the investigator is not following his protocol and is conducting research in a manner that unnecessarily produces pain and suffering, I can have his protocol shut down until he complies. I can't do anything about the use of animals in research, but I do believe that we on the IACUC work very hard to ensure that researchers do not take animal lives for granted, and that they do not suffer excessively.

Susan M. Harnett has a BS in Mass Communications/Film from Emerson College, Boston, MA; a Master's of Library Service from the School of Communication and Information, Rutgers University, New Brunswick, NJ. She is Medical Information Services Librarian at Borland Library, University of Florida, Jacksonville, FL.

That feeling of "how do I fit into this group" easily bridges the reader from the Harnett narrative above to Megan Sheffield's, the last narrative of the final chapter of this book. Sheffield feels she has taken an unusual path into librarianship; her academic background (including and other than in librarianship) has much in common with the other voices in this chapter. Her steps toward librarianship may be more accidental, yet her steps onto her university's IACUC are familiar—a retirement or vacancy left an open spot, which also means a window of opportunity and for conversation.

The key thing here is Sheffield's mind remaining open to possibilities for engagement and growth, and the past experiences of the IACUC with having a librarian on board left them understanding a librarian's capacity to fill that role. That said, as she notes, she initially served in a more consultation capacity, rather than as a true committee member, and after professional development and continued participation as a consultant, the IACUC asked Sheffield to join the committee in earnest. Perhaps more unusual among this group, Sheffield became the data services librarian *after* spending several years as the science librarian and *after* being on the IACUC over that time. Her work connects her directly with the Office of Research Compliance (OCR) and has given her expertise with data-sharing policies relevant to her university and its stakeholders, as well as with key developing topic areas like predatory publishing and open-access initiatives.

BECOMING A LIBRARIAN AND AN IACUC MEMBER: LESSONS IN SKILL SETS AND TRANSITIONS

I had an unusual path to librarianship. I initially got a BS and MS in biology. I decided halfway through my MS that I wanted a change of career path, and my part-time job working at the university library reference desk was great, so I decided to get my MLIS. After I got a tenure-track job (initially as a science librarian), I was looking for committee work and was told that a librarian had just retired from the IACUC, so I started talking with that committee. At first, I wasn't actually on the committee, but I was mentioned in the documentation to encourage researchers to contact me for help with their literature reviews. I took a two-day intensive at the National Agricultural Library in Beltsville, Maryland, on searching the literature for animal welfare requirements. I brought this knowledge back to my campus and started giving lunchtime drop-in sessions (sponsored by our Office of Research Compliance [ORC]) about that and other topics. After a couple of years, I was asked to join the IACUC in the "non-scientist" role, so I attend all meetings and vote on new protocols. After working at the library as a science librarian for six years and achieving tenure, I transitioned to a new role as data services librarian, which is a better fit for the type of work I found myself doing with science researchers.

I've found data librarianship to have several things in common with serving on the IACUC. There's a similar emphasis placed on research ethics in both areas, and I've served as an educator to researchers for both. One key area of overlap is RCR training; this "Responsible Conduct of Research" training is mandated by federal grants and has a list of specific areas of training such as research ethics, data management, and animal welfare requirements. By collaborating with the IACUC and Office of Research Compliance, I can do my job to educate researchers about issues regarding their data, the researchers can meet their training requirements, and the ORC and IACUC get help providing a full slate of RCR opportunities. I've also found there are other unique areas of overlap, for example, Ag Data Commons. As a science librarian at a land grant university with strong agriculture and animal science programs, I have to be particularly well versed in the United States Department of Agriculture's data-sharing policies. I can serve the animal research community (and therefore IACUC) by providing training on depositing or finding data in Ag Data Commons.

I've transitioned to a new role in the library as data services librarian, but I still have a close relationship with the Office of Research Compliance. Our missions overlap in a few key areas like data sharing, concerns about authorship and predatory publishing, and open access publishing. Although I could technically step back from IACUC at this point, as we have other science librarians that could fill the role, my particular background and training make me a good fit for it and, honestly, it's one of the most rewarding committees I serve on, so I am in no rush to give it up. I'd strongly encourage other librarians that might be interested in this niche to give it a shot!

Megan Sheffield has a BS and an MS in biology from Clemson University and an MLIS in library science from the University of Maryland–College Park. She is the data services librarian and associate librarian liaison to the College of Agriculture, Forestry, and Life Sciences as well as the Biological Sciences Department at Clemson University Libraries.

This chapter has addressed a variety of voices, pathways, skill sets, (ad) ventures, fits, motivations, and opportunities for academic librarians to connect themselves with their IRBs and IACUCs through matching skills and goals, but also through displaying respect for research, its goals, its processes, and its participants. These are values librarians often share and support, even though the support can many times be hidden. Librarians are well positioned to play important roles, but these capacities must be supported through networking, professional development, campus positioning of the library, and a certain *carpe diem* attitude to seeking—and seeing—opportunities. Not sure if there's a place for you, or a colleague, on your institution's IRB or IACUC? This chapter and, indeed, this whole book, offers not only information and guidance, but also contacts. We must often provide proof of concept and our voices, separately as distinct individuals and together as a likeminded cohort, all serve in that capacity.

BIBLIOGRAPHY

Berg, Irwin A. "The Use of Human Subjects in Psychological Research." *American Psychologist* 9, no. 3 (1954): 108–111. https://doi.org/10.1037/h006338.

Taylor, Judith and Matthew Patterson. "Autonomy and Compliance: How Qualitative Sociologists Respond to Institutional Ethical Oversight." *Qualitative Sociology* 33 (2010): 161–183. DOI 10.1007/s11133-010-9148-y.

IRB and IACUC Resources and Readings to Know and Explore

SEMINAL DOCUMENTS AND READINGS

Belmont Report (https://www.hhs.gov/ohrp/regulations-and-policy/belmont-report/index.html)

Declaration of Helsinki (https://www.wma.net/policies-post/wma-declaration-of-helsinki-ethical-principles-for-medical-research-involving-human-subjects/)

Good Clinical Practice Guidelines (https://ichgcp.net/)

Guide for Care and Use of Laboratory Animals (https://www.aaalac.org/the-guide/)

Nuremberg Code (https://history.nih.gov/spaces/flyingpdf/pdfpageexport.action?pageId=30736929)

PHS Policy on Humane Care and Use of Laboratory Animals (https://olaw.nih.gov/policies-laws/phs-policy.htm)

The Principles of Humane Experimental Technique by Russell and Burch (https://caat.jhsph.edu/principles/the-principles-of-humane-experimental-technique)

Reducing Administrative Burden for Researchers: Animal Care and Use in Research (August 2019 report; https://olaw.nih.gov/sites/default/files/21CCA_final_report.pdf)

Universal Declaration of Human Rights (https://www.un.org/sites/un2.un.org/files/udhr.pdf)

FEDERAL LAWS AND OFFICES

Code of Federal Regulations (C.F.R.) (https://www.ecfr.gov/cgi-bin/ECFR?page=browse)—Common Rule (45 C.F.R. 46) (https://www.hhs.gov/ohrp/regulations-and-policy/regulations/45-cfr-46/index.html) Protection of Human Subjects (21 C.F.R. 50)—(https://www.ecfr.gov/cgi-bin/text-idx?SID=9c8b6a0067560c418

bab3b3ba9a2610c&mc=true&node=pt21.1.50&rgn=div5) Institutional Review Boards (21 C.F.R. 56) – https://www.ecfr.gov/cgi-bin/text-idx?SID=9c8b6a00675 60c418bab3b3ba9a2610c&mc=true&node=pt21.1.56&rgn=div5)

Clinical Trials (https://clinicaltrials.gov/)

Department of Health and Human Services (DHHS) (https://www.hhs.gov/)—Office of Human Research Protections (OHRP; replaced Office of Protection from Research Risks in 2000; https://www.hhs.gov/ohrp/) OHRP special reports: https://www.hhs.gov/ohrp/regulations-and-policy/belmont-report/access-other-reports-by-the-national-commission/index.html

Food and Drug Administration (FDA) (https://www.fda.gov/)—Good Clinical Practice (https://www.fda.gov/about-fda/center-drug-evaluation-and-research-cder/good-clinical-practice)

National Institutes of Health (NIH) (https://www.nih.gov/)—National Institute of Environmental Health Sciences (https://www.niehs.nih.gov/), including ALTBIB—Alternatives to Animal Testing (https://ntp.niehs.nih.gov/whatwestudy/niceatm/altbib/index.html)

Office of Extramural Research (OER) (https://grants.nih.gov/aboutoer/intro2oer.htm)

Office of Laboratory Animal Welfare (OLAW) (https://olaw.nih.gov/home.htm)

National Science Foundation (NSF) (https://www.nsf.gov/)

United States Code (U.S.C.) (https://uscode.house.gov/)—Animal Welfare Act (AWA) (7 U.S.C. § 2131 et seq)

The Public Health and Welfare Act (PHWA) (42 U.S.C. 42)

USDA (https://www.usda.gov/)—Animal and Plant Health Inspection Service (APHIS) (https://www.aphis.usda.gov/aphis/home/) Animal Welfare Information Center (AWIC) (https://www.nal.usda.gov/awic)

ORGANIZATIONS, TRAININGS, AND CONFERENCES

Association for Assessment and Accreditation of Laboratory Animal Care (AAALAC) (https://www.aaalac.org/)

Collaborative Institutional Training Institute (CITI) (https://about.citiprogram.org/en/homepage/)

Council for International Organizations of Medical Sciences (CIOMS) (https://cioms.ch/)

International Ethical Guidelines for Biomedical Research Involving Human Subjects (https://cioms.ch/wp-content/uploads/2021/03/International_Ethical_Guidelines_for_Biomedical_Research_Involving_Human_Subjects_2002.pdf)

European 3Rs Centers (https://ec.europa.eu/jrc/en/eurl/ecvam/knowledge-sharing-3rs/knowledge-networks/eu-3rs-centres)

Institute for Laboratory Animal Research (ILAR) (https://www.nationalacademies.org/ilar/institute-for-laboratory-animal-research)

Johns Hopkins Bloomberg School of Public Health (JHSPH) Center for Alternatives to Animal Testing (CAAT) (https://caat.jhsph.edu/)

Massachusetts Society for Medical Research (MSMR) (https://msmr.org/)
Norwegian Inventory of Alternatives (NORINA) (https://norecopa.no/norina)
Public Responsibility in Medicine and Research (PRIM&R) (https://www.primr.org/)
United Nations Educational, Scientific, and Cultural Organization (UNESCO) (https://en.unesco.org/)

Index

Editors and Contributors

EDITORS

Susan M. Harnett, MLS, AHIP-D, is the Medical Information Services librarian at Borland Library, University of Florida, Jacksonville, Florida. She has been the nonscientific member of the Institutional Animal Care and Use Committee (IACUC) at the University of Florida since 2016; she has also served as a voting member of the institutional review board at Naval Medical Center, Portsmouth, Virginia, and as an ex-officio member of both the institutional review board (IRB) and IACUC at Eastern Virginia Medical School, Norfolk, Virginia.

Laureen P. Cantwell, MSLIS, is the head of access services and outreach at Colorado Mesa University's Tomlinson Library. She oversees their Checkout + Reserves service point and staff, as well as their resource sharing/ILL staff. Laureen is currently pursuing a PhD in information science from University at Buffalo–SUNY. She also coedited *Memphis Noir* (2015) and has published book chapters and articles on topics ranging from librarians on institutional review boards (IRBs) to Massive Open Online Courses (MOOCs) to curbside pickup services, from chat reference to library printing to digital badging, and more.

CONTRIBUTORS

Eric D. Albright, AM MA, AHIP, is the director of the Hirsh Health Sciences Library of Tufts University. Prior to coming to Tufts he was at Duke University's Medical Center Library and Northwestern University's Galter

Health Sciences Library. He has his library degree from the University of Chicago and a masters in bioethics and health policy from Loyola University. He is currently on the institutional review board (IRB) of Tufts where he started in 2012 and is also on the IRB of Salem State University where he started in 2008. He is active in the Medical Library Association, the American Library Association, and the Association of Academic Health Science Libraries as well as the author or coauthor of many reviews and articles.

Kristine M. Alpi is university librarian and associate professor, Oregon Health & Science University and adjunct assistant professor of population health and pathobiology at North Carolina State University (NCSU). She holds a masters of library science from Indiana University, a masters of public health from Hunter College, City University of New York, and a PhD in educational research and policy analysis from NCSU. A former member of the NLM Biomedical Library and Informatics Review Committee Study Section, Alpi continues as a reviewer and chair for special emphasis panels. A distinguished member of the Academy of Health Information Professionals of the Medical Library Association (MLA), Alpi's MLA activities include previously chairing the Research and Public Health/Health Administration sections, serving on the board of directors, and as president for 2020–2021.

Karen D. Barton is the biomedical research liaison Librarian at Duke University Medical Center Library and Archives, and she is a senior member of the Academy of Health Information Professionals (AHIP). She works closely with the research community to support the research lifecycle through advanced and systematic literature searching, NIH Public Access Policy compliance efforts, research impact services, and other projects. Karen also serves as a liaison to and member of the Institutional Animal Care and Use Committee (IACUC), a dual role in which she assists with searches for animal alternatives, conducts animal facility inspections, and serves as a reviewer of research protocol amendments. Karen earned an MSLIS from the University of Illinois at Urbana–Champaign.

Everly D. Brown is the head of information services at the Health Sciences and Human Services Library, University of Maryland, Baltimore. She ensures excellent customer service at the public service desk, leads communication and promotional initiatives for the library, provides expertise on access issues to the library's resources, and develops and supports innovative services to advance research and scholarship at UMB, such as the HS/HSL's makerspace, institutional review board consent form review service, and poster printing service. She received her MLIS degree from the University of Texas at Austin.

Robin E. Champieux is the director of education, research and clinical outreach and associate professor at Oregon Health & Science University (OHSU), where she leads the library's education, scholarly communication, and information services. Her research is focused on enabling the creation, reproducibility, accessibility, and impact of digital scientific materials. She is the cofounder of the Metrics Toolkit and Awesome Libraries. Robin was a 2018 NLM/AAHSL Leadership Fellow and is a graduate of the Harvard Leadership Institute for Academic Librarians. Robin is also a member of the OHSU site team for the Center for Data to Health (CD2H), where she contributes to efforts to promote and facilitate data reuse and interoperability, tool sharing, informatics fluency, and collaboration.

Melissa C. Funaro, MLS, MS, is a clinical librarian at Yale's Harvey Cushing/John Hay Whitney Medical Library. She earned her master of library science from Southern Connecticut State University, and a master of science degree in horticulture from West Virginia University. Melissa serves as a member of Yale's Institutional Animal Care and Use Committee (IACUC) and policy subcommittee. She provides research and instruction supports to several departments at the Yale School of Medicine including psychiatry, pediatrics, and the child study center.

Emily F. Gorman is a research, education and outreach librarian and the liaison to the School of Pharmacy from the Health Sciences and Human Services Library at the University of Maryland, Baltimore (UMB). She received her master of library and information science (MLIS) degree from the University of Pittsburgh. Prior to arriving at UMB in April 2017 she served as the health and natural sciences librarian at the Loyola Notre Dame Library in Baltimore, where she developed a particular interest in health sciences librarianship while working with the schools of pharmacy and nursing from Notre Dame of Maryland University. Her main areas of research are instructional techniques and alternative metrics for measuring research impact.

Nathan Hall is a professor at Virginia Tech University Libraries where he is director of digital imaging and preservation services. He serves on Virginia Tech's institutional review board (IRB) and has also contributed to university and consortial research and data governance through various committee roles. As a former chair of ACRL's Research and Scholarly Environments Committee, he contributed to the development of ACRL's "Open and Equitable Scholarly Communications: Creating a More Inclusive Future," a research agenda that advocates for more open research infrastructure and policy.

Andrea M. Harrow, MLIS, AHIP-D, is a solo, medical librarian at PIH Health Good Samaritan Hospital, Los Angeles, where she has worked for the past seventeen years. She earned her master's in information services management from University of North London, now London Metropolitan University in Finsbury Park, London. Previously, she was the annuals librarian for the British Film Institute National Library.

Wendy Highby was a paralegal for over a decade, then changed professions and became a librarian. She worked on a bookmobile and at a historical archive before finding her niche in academic librarianship. She is currently social sciences librarian and professor at the University of Northern Colorado. Her scholarship covers a variety of topics, including popular culture (*The Colbert Report*'s representation of authors), creative copyright pedagogy, and campus environmentalism. She coauthored the book *Beyond Accommodation: Creating an Inclusive Workplace for Disabled Library Workers* (2020). In 2021 she completes eleven years of service on her university's institutional review board.

Brian Jackson is an associate professor and data librarian at Mount Royal University Library in Calgary, Alberta. He leads library research data management services, including oversight for the institution's data repository, and supports the responsible management of research data through service on the university's human research ethics board. In addition to data services, Brian provides research and information literacy support to faculty and students in several programs, including earth and environmental sciences, policy studies, and economics. Brian's research focuses on research data management policies and ethical practice.

Andrea C. Kepsel MLS, AHIP, is a health sciences educational technology librarian at Michigan State University (MSU), where she is a liaison to the College of Veterinary Medicine, the Department of Animal Sciences, the Biomedical Laboratory Diagnostics Program, the Department of Biomedical Engineering, and the Department of Food Science and Human Nutrition. She has a bachelor's of science from Michigan State University and a masters of library and information science from Wayne State University. Prior to becoming a librarian, Andrea conducted research in molecular biology and biomedical engineering. In addition to her liaison and reference responsibilities, Andrea teaches evidence-based veterinary medicine in the DVM curriculum and has coauthored an article in the *Journal of the Medical Library Association* about how veterinary students seek out and evaluate scientific literature. She is a scientific voting member of MSU's Institutional Animal Care and Use Committee.

Rebekah S. Miller is a research and instruction librarian and liaison to the School of Nursing and the Institutional Animal Care and Use Committee (IACUC) at the Health Sciences Library System, University of Pittsburgh. She received her MLIS from the University of Pittsburgh. Her research interests include critical pedagogy, research ethics, and retractions.

Michele Nance works as a reference associate at the University of Maryland, Baltimore's Health Sciences and Human Services Library. She holds a master's degree in professional writing from Towson University and a bachelor's degree in humanities (comparative literature) from Johns Hopkins University.

Daureen Nesdill graduated from SUNY Downstate Medical School, College of Health Science with a major in laboratory animal science and spent the next five years working as an animal technician. Allergies changed her career focus and Daureen earned a MS in ecology and an MLIS before starting a position as a STEM librarian at the Marriott Library, University of Utah in 2002. She first volunteered to serve on the Institutional Animal Care and Use Committee (IACUC) in 2004 and remains an active member. Daureen's medical and ecological research experience and her interest in data management led to her being named the data management librarian in 2006.

Kate Nyhan, MLS, is a research and educational librarian at Yale's Harvey Cushing / John Hay Whitney Medical Library and a lecturer in environmental health sciences at Yale School of Public Health. She is pursuing a post-master's certificate as an inter-professional informationist at Simmons University, funded by the Institute of Museum and Library Services. Kate is active in the Medical Library Association's public health / health administration caucus. Her research interests include reporting guidelines, reproducibility, and research waste.

Melissa A. Ratajeski an assistant director for data and publishing services and liaison to the Institutional Animal Care and Use Committee (IACUC) at the Health Sciences Library System, University of Pittsburgh. In this role, she provides support and training for researchers at each stage of the data lifecycle. Before receiving her master's in library and information science, Melissa was a research coordinator within the Department of Neurobiology/ Center for the Neural Basis of Cognition, at the University of Pittsburgh School of Medicine.

Esther May Sarino, MLIS, has a master's in library and information science from Drexel University. She currently works for Eastern Virginia Medical

School as the reference services coordinator for the Brickell Medical Sciences Library. Her job as coordinator encompasses overseeing the reference and clinical medical librarian services, including services to the institutional review board.

Marcelle Savoy is the medical librarian at Lincoln Memorial University serving the DeBusk College of Osteopathic Medicine. Until recently, she was also the interim veterinary medicine librarian and served as a voting member on the Institutional Animal Care and Use Committee (IACUC). She currently rotated off as an alternate member after a new veterinary medicine/access services librarian was hired. Her background is in microbiology, and she taught high school biology and chemistry for twenty years before becoming a medical librarian. Her interest areas include all types of reviews (narrative, scoping, systematic, etc.), evidence-based medicine, and bioinformatics. She has coauthored several narrative and literature reviews, which have recently been published.

Megan Sheffield received her BS and MS in biology from Clemson University and her MLIS from the University of Maryland College Park. She has performed research with vertebrate animals and served as a science reference librarian, and currently works as the data services librarian at Clemson University. Her research interests include research compliance, scholarly communications trends, and predatory publishing.

Tracy C. Shields, MSIS, is the reference medical librarian at Naval Medical Center Portsmouth in Virginia, where she serves as a nonscientist member on the institutional review board (IRB) and the Institutional Animal Care and Use Committee (IACUC). With over fifteen years of experience in medical librarianship, she provides in-depth reference and searching in support of graduate medical education, evidence-based practice, systematic reviews, and the research lifecycle. Tracy's professional passions include searching, discovering gray literature (especially for public health, disaster preparedness, and humanitarian aid), scholarly communication, and advocating for health equity. She is credentialed as a senior member of the Academy of Health Information Professionals by the Medical Library Association and has a specialization in disaster information.

John Sisson has been the research librarian for biology at the University of California, Irvine since 1990. He has a bachelor's degree in biology from University of California, San Diego, a masters degree in zoology from the University of Maine, Orono, and a masters of library and information science from the University of Hawaii, Manoa. He has been a member of the

University of California, Irvine Institutional Animal Care and Use Committee (IACUC) since 2004.

Faythe Thurman is the access services librarian at Lincoln Memorial University, where she has worked since 2018. As liaison to the College of Veterinary Medicine and the Veterinary Health Science and Technology program, she collaborates with faculty on research projects and integrating information literacy instruction into the curriculum. In her position, Faythe serves as a voting member of Lincoln Memorial University's Institutional Animal Care and Use Committee (IACUC). She earned an MS in library and information science from Clarion University and an MA and BA from Shippensburg University

Marijane K. White is the data and research engagement librarian and assistant professor at Oregon Health & Science University (OHSU), where she assists members of the OHSU community with a wide variety of data needs, measuring research impact, making research reproducible, selecting computational tools, and answering copyright questions. In addition to her library degree, she holds an undergraduate degree in computer engineering and has fifteen years of experience in various software industry roles, a background that has proven useful when advising researchers about making the computational aspects of their research reproducible. In 2019, she co-facilitated the Rigor and Reproducibility nanocourse for OHSU's Program in Enhanced Research Training (PERT) with metabolomics researcher Jean-Phillipe Gourdine. Marijane focuses on using ontologies to create semantic data models of research profiles, research networks, research contributions, healthcare entities, and the landscape of rubrics for implementing the FAIR Data Principles.